Complex Integration and Cauchy's Theorem

COMPLEX INTEGRATION
AND
CAUCHY'S THEOREM

by

G. N. WATSON, M.A.

Fellow of Trinity College, Cambridge

Cambridge:
at the University Press
1914

PREFACE

MY object in writing this Tract was to collect into a single volume those propositions which are employed in the course of a rigorous proof of Cauchy's theorem, together with a brief account of some of the applications of the theorem to the evaluation of definite integrals.

My endeavour has been to place the whole theory on a definitely arithmetical basis without appealing to geometrical intuitions. With that end in view, it seemed necessary to include an account of various propositions of *Analysis Situs*, on which depends the proof of the theorem in its most general form. In proving these propositions, I have followed the general course of a memoir by Ames; my indebtedness to it and to the textbooks on Analysis by Goursat and by de la Vallée Poussin will be obvious to those who are acquainted with those works.

I must express my gratitude to Mr Hardy for his valuable criticisms and advice; my thanks are also due to Mr Littlewood and to Mr H. Townshend, B.A., Scholar of Trinity College, for their kindness in reading the proofs.

G. N. W.

TRINITY COLLEGE,
February 1914.

CONTENTS

INTRODUCTION

1. Throughout the tract, wherever it has seemed advisable, for the sake of clearness and brevity, to use the language of geometry, I have not hesitated to do so; but the reader should convince himself that all the arguments employed in Chapters I—IV are really arithmetical arguments, and are not based on geometrical intuitions. Thus, no use is made of the geometrical conception of an angle; when it is necessary to define an angle in Chapter I, a purely analytical definition is given. The fundamental theorems of the arithmetical theory of limits are assumed.

A number of obvious theorems are implicitly left to the reader; e.g. that a circle is a 'simple' curve (the coordinates of any point on $x^2+y^2=1$ may be written $x=\cos t$, $y=\sin t$, $0 \leqslant t \leqslant 2\pi$); that two 'simple' curves with a common end-point, but with no other common point, together form one 'simple' curve; and several others of a like nature.

It is to be noted that almost all the difficulties, which arise in those problems of *Analysis Situs* which are discussed in Chapter I, disappear if the curves which are employed in the following chapters are restricted to be straight lines or circles. This fact is of some practical importance, since, in applications of Cauchy's Theorem, it is usually possible to employ only straight lines and circular arcs as contours of integration.

2. Notation. If z be a complex number, we shall invariably write

$$z = x + iy,$$

where x and y are real; with this definition of x and y, we write[1]

$$x = R(z), \quad y = I(z).$$

If a complex number be denoted by z with some suffix, its real and imaginary parts will be denoted by x and y, respectively, with the same suffix; e.g.

$$z_n = x_n + iy_n;$$

[1] The symbols R and I are read 'real part of' and 'imaginary part of' respectively.

further, if ζ be a complex number, we write

$$R(\zeta) = \xi, \quad I(\zeta) = \eta.$$

Definitions. *Point.* A 'point' is a value of the complex variable, z; it is therefore determined by a complex number, z, or by two real numbers (x, y). It is represented geometrically by means of the Argand diagram.

Variation and Limited Variation[2]. If $f(x)$ be a function of a real variable x defined when $a \leqslant x \leqslant b$ and if numbers $x_1, x_2, \ldots x_n$ be chosen such that $a \leqslant x_1 \leqslant x_2 \ldots \leqslant x_n \leqslant b$, then the sum

$$|f(x_1)-f(a)|+|f(x_2)-f(x_1)|+|f(x_3)-f(x_2)|+\ldots+|f(b)-f(x_n)|$$

is called the *variation of $f(x)$ for the set of values $a, x_1, x_2, \ldots x_n, b$.* If for every choice of $x_1, x_2, \ldots x_n$, the variation is always less than some finite number λ (independent of n), $f(x)$ is said to have *limited variation* in the interval a to b; and the upper limit of the variation is called the *total variation* in the interval.

[The notion of the variation of $f(x)$ in an interval a to b is very much more fundamental than that of the length of the curve $y=f(x)$; and throughout the tract propositions will be proved by making use of the notion of *variation* and not of the notion of *length*.]

[2] Jordan, *Cours d'Analyse*, §§ 105 *et seq*.

CHAPTER I

ANALYSIS SITUS

§ 3. Problems of *Analysis situs* to be discussed.—§ 4. Definitions.—§ 5. Properties of continua.—§ 6. Theorems concerning the order of a point.—§ 7. Main theorem; a regular closed curve has an interior and an exterior.—§ 8. Miscellaneous theorems; definitions of counterclockwise and orientation.

3. The object of the present chapter is to give formal analytical proofs of various theorems of which simple cases seem more or less obvious from geometrical considerations. It is convenient to summarise, for purposes of reference, the general course of the theorems which will be proved:

A *simple curve* is determined by the equations $x=x(t), y=y(t)$ (where t varies from t_0 to T), the functions $x(t)$, $y(t)$ being continuous, and the curve has no double points save (possibly) its end points; if these coincide, the curve is said to be *closed*. The *order* of a point Q with respect to a closed curve is defined to be n, where $2\pi n$ is the amount by which the angle between QP and Ox increases as P describes the curve once. It is then shewn that points in the plane, not on the curve, can be divided into two sets; points of the first set have order ± 1 with respect to the curve, points of the second set have order zero; the first set is called the interior of the curve, and the second the exterior. It is shewn that *every* simple curve joining an interior point to an exterior point must meet the given curve, but that simple curves can be drawn, joining any two interior points (or exterior points), which have no point in common with the given curve. It is, of course, not obvious that a closed curve (defined as a curve with coincident end points) divides the plane into two regions possessing these properties.

It is then possible to distinguish the direction in which P describes the curve (viz. counterclockwise or clockwise); the criterion which determines the direction is the sign of the order of an interior point.

The investigation just summarised is that due to Ames[1]; the analysis which will be given follows his memoir closely. Other proofs that a closed curve

[1] Ames, *American Journal of Mathematics*, Vol. XXVII. (1905), pp. 343–380.

possesses an interior and an exterior have been given by Jordan[2], Schoenflies[3], Bliss[4], and de la Vallée Poussin[5]. It has been pointed out that Jordan's proof is incomplete, as it assumes that the theorem is true for closed polygons; the other proofs mentioned are of less fundamental character than that of Ames.

4. DEFINITIONS. A *simple curve* joining two points z_0 and Z is defined as follows:

Let[6]
$$x = x(t), \quad y = y(t),$$
where $x(t)$, $y(t)$ are continuous one-valued functions of a real parameter t for all values of t such that[7] $t_0 \leq t \leq T$; the functions $x(t)$, $y(t)$ are such that they do not assume the same pair of values for any two different values of t in the range $t_0 < t < T$, and
$$z_0 = x(t_0) + iy(t_0), \quad Z = x(T) + iy(T).$$
Then we say that the *set of points* (x, y), determined by the set of values of t for which $t_0 \leq t \leq T$, is a *simple curve* joining the points z_0 and Z. If $z_0 = Z$, the simple curve is said to be *closed*[8].

To render the notation as simple as possible, if the parameter of any particular point on the curve be called t with some suffix, the complex coordinate of that point will always be called z with the same suffix; thus, if
$$t_0 \leq t_r^{(n)} \leq T,$$
we write
$$z_r^{(n)} = x(t_r^{(n)}) + iy(t_r^{(n)}) = x_r^{(n)} + iy_r^{(n)}.$$

Regular curves. A simple curve is said to be *regular*[9], if it can be divided into a *finite* number of parts, say at the points whose parameters are $t_1, t_2, \ldots t_m$ where $t_0 \leq t_1 \leq t_2 \leq \ldots \leq t_m \leq T$, such that when

[2] Jordan, *Cours d'Analyse* (1893), Vol. I. §§ 96–103.

[3] Schoenflies, *Göttingen Nachrichten*, Math.-Phys. Kl. (1896), p. 79.

[4] Bliss, *American Bulletin*, Vol. X. (1904), p. 398.

[5] de la Vallée Poussin, *Cours d'Analyse* (1914), Vol. I. §§ 342–344.

[6] The use of x, y in two senses, as coordinates and as functional symbols, simplifies the notation.

[7] We can always choose such a parameter, t, that $t_0 < T$; for if this inequality were not satisfied, we should put $t = -t'$ and work with the parameter t'.

[8] The word 'closed' except in the phrase 'closed curve' is used in a different sense; a *closed* set of points is a set which contains all the limiting points of the set; an open set is a set which is not a closed set.

[9] We do not follow Ames in assuming that $x(t)$, $y(t)$ possess derivatives with regard to t.

$t_{r-1} \leqslant t \leqslant t_r$, the relation between x and y given by the equations $x = x(t)$, $y = y(t)$ is equivalent to an equation $y = f(x)$ *or else* $x = \phi(y)$, where f *or* ϕ denotes a continuous one-valued function of its argument, and r takes in turn the values 1, 2, . . $m+1$, while $t_{m+1} = T$.

It is easy to see that a chain of a finite number of curves, given by the equations

$$\left.\begin{array}{ll} y=f_1(x), & a_1 \leqslant x \leqslant a_2 \\ x=f_2(y), & b_2 \leqslant y \leqslant b_3 \\ y=f_3(x), & a_3 \leqslant x \leqslant a_4 \\ \dots\dots\dots & \dots\dots\dots \end{array}\right\} \quad \dots \quad \dots\dots\dots \text{(A)}$$

(where $b_2 = f_1(a_2)$, $a_3 = f_2(b_3)$, ... and f_1, f_2, ... are continuous one-valued functions of their arguments), forms a simple curve, if the chain has no double points; for we may choose a parameter t, such that

$$x=t, \quad y=f_1(t), \quad a_1 \leqslant t \leqslant a_2;$$

$$x=f_2(b_2-a_2+t), \quad y=b_2-a_2+t, \quad a_2 \leqslant t \leqslant a_2-b_2+b_3;$$

$$x=a+t, \quad y=f_3(a+t), \quad a_3-a \leqslant t \leqslant a_4-a, \quad a=a_3-a_2+b_2-b_3,$$

$$\dots\dots\dots\dots\dots\dots\dots\dots\dots\dots\dots\dots\dots\dots\dots$$

If some of the inequalities in equations (A) be reversed, it is possible to shew in the same manner that the chain forms a simple curve.

Elementary curves. Each of the two curves whose equations are (i) $y = f(x)$, $(x_0 \leqslant x \leqslant x_1)$ and (ii) $x = \phi(y)$, $(y_0 \leqslant y \leqslant y_1)$, where f and ϕ denote one-valued continuous functions of their respective arguments, is called an *elementary* curve.

Primitive period. In the case of a closed simple curve let $\omega = T - t_0$; we define the functions $x(t)$, $y(t)$ for *all* real values of t by the relations

$$x(t+n\omega) = x(t), \quad y(t+n\omega) = y(t),$$

where n is any integer; ω is called the primitive period of the pair of functions $x(t)$, $y(t)$.

Angles. If z_0, z_1 be the complex coordinates of two distinct points P_0, P_1, we say that 'P_0P_1 makes an angle θ with the axis of x' if θ satisfies *both* the equations[10]

$$\cos\theta = \kappa(x_1 - x_0), \quad \sin\theta = \kappa(y_1 - y_0),$$

where κ is the positive number $\{(x_1-x_0)^2 + (y_1-y_0)^2\}^{-\frac{1}{2}}$. This pair of equations has an infinite number of solutions such that if θ, θ' be any

[10] It is supposed that the sine and cosine are defined by the method indicated by Bromwich, *Theory of Infinite Series*, § 60, (2); it is easy to deduce the statements made concerning the solutions of the two equations in question.

two different solutions, then $(\theta'-\theta)/2\pi$ is an integer, positive or negative.

Order of a point. Let a regular closed curve be defined by the equations $x=x(t)$, $y=y(t)$, $(t_0 \leqslant t \leqslant T)$ and let ω be the primitive period of $x(t)$, $y(t)$. Let Q be a point *not* on the curve and let P be the point on the curve whose parameter is t. Let $\theta(t)$ be the angle which QP makes with the axis of x; since every branch of arc cos $\{\kappa(x_1-x_0)\}$ and of arc sin $\{\kappa(y_1-y_0)\}$ is a continuous function of t, it is possible to choose $\theta(t)$ so that $\theta(t)$ is a continuous function of t reducing to a definite number $\theta(t_0)$ when t equals t_0. The points represented by the parameters t and $t+\omega$ are the same, and hence $\theta(t)$, $\theta(t+\omega)$ are two of the values of the angle which QP makes with the axis of x; therefore

$$\theta(t+\omega)-\theta(t)=2n\pi,$$

where n is an integer; n is called the *order* of Q with respect to the curve. To shew that n depends only on Q and not on the particular point, P, taken on the curve, let t vary continuously, then $\theta(t)$, $\theta(t+\omega)$ vary continuously; but since n is an integer n can only vary *per saltus*. Hence n is constant[11].

5. CONTINUA. A two-dimensional continuum is a set of points such that (i) if z_0 be the complex coordinate of any point of the set, a positive number δ can be found such that all points whose complex coordinates satisfy the condition $|z-z_0|<\delta$ belong to the set; δ is a number depending on z_0, (ii) any two points of the set can be joined by a simple curve such that all points of it belong to the set.

Example. The points such that $|z|<1$ form a continuum.

[11] This argument *really* assumes what is known as Goursat's lemma (see § 12) for functions of a real variable. It is proved by Bromwich, *Theory of Infinite Series*, p. 394, example 18, that if an interval has the property that round every point P of the interval we can mark off a sub-interval such that a certain inequality denoted by $\{Q, P\}$ is satisfied for every point Q of the sub-interval, then we can divide the whole interval into a *finite* number of closed parts such that each part contains at least one point P_1 such that the inequality $\{Q, P_1\}$ is satisfied for all points Q of the part in which P_1 lies.

In the case under consideration, we have a function, $\phi(t)=\theta(t+\omega)-\theta(t)$, of t, which is given continuous; the inequality is therefore $|\phi(t)-\phi(t')|<\epsilon$, where ϵ is an arbitrary positive number; by the lemma, taking $\epsilon<2\pi$, we can divide the range of values of t into a finite number of parts in each of which $|\phi(t)-\phi(t_1)|<2\pi$ and is therefore zero; $\phi(t)$ is therefore constant throughout each part and is therefore constant throughout the sub-interval.

Neighbourhood, Near. If a point ζ be connected with a set of points in such a way that a sequence (z_n), consisting of points of the set, can be chosen such that ζ is a limiting point of the sequence, then the point ζ is said to have points of the set *in its neighbourhood.*

The statement 'all points *sufficiently near* a point ζ have a certain property' means that a positive number h exists such that all points z satisfying the inequality $|z-\zeta|<h$ have the property.

Boundaries, Interior and Exterior Points. Any point of a continuum is called an *interior* point. A point is said to be a *boundary point* if it is not a point of the continuum, but has points of the continuum in its neighbourhood.

A point z_0, such that $|z_0|=1$, is a boundary point of the continuum defined by $|z|<1$

A point which is not an interior point or a boundary point is called an *exterior* point

If (z_n) be a sequence of points belonging to a continuum, then, if this sequence has a limiting point ζ, the point ζ is either an interior point or a boundary point; for, even if ζ is not an interior point, it has points of the continuum in its neighbourhood, viz. points of the sequence, and is therefore a boundary point.

All points sufficiently near an exterior point are exterior points; for let z_0 be an exterior point; then, if *no* positive number h exists such that *all* points satisfying the inequality $|z-z_0|<h$ are exterior points, it is possible to find a sequence (ζ_n) such that ζ_n is an interior point or a boundary point and $|\zeta_n-z_0|<2^{-n}$; and, whether ζ_n is an interior point or a boundary point, it is possible to find an *interior* point ζ_n' such that $|\zeta_n'-\zeta_n|<2^{-n}$; so that $|\zeta_n'-z_0|<2^{1-n}$, and z_0 is the limiting point of the sequence ζ_n'; therefore z_0 is an interior point or a boundary point; this is contrary to hypothesis; therefore, corresponding to any particular point z_0, a number h exists. The theorem is therefore proved.

A continuum is called an[12] *open region*, a continuum with its boundary is a *closed region.*

Example. Let S be a set of points $z\ (=x+iy)$ defined by the relations

$$x_0<x<x_1, \quad y=f(x)+r \quad \text{.........................(1)},$$

where f is one-valued and continuous, r takes all values such that $0<r<k$, and k is constant. Then the set of points S forms a continuum.

[12] See note 8 on p. 4.

Let z' be any point of S, so that

$$x_0 < x' < x_1, \quad y' = f(x') + r', \quad \text{where } 0 < r' < k.$$

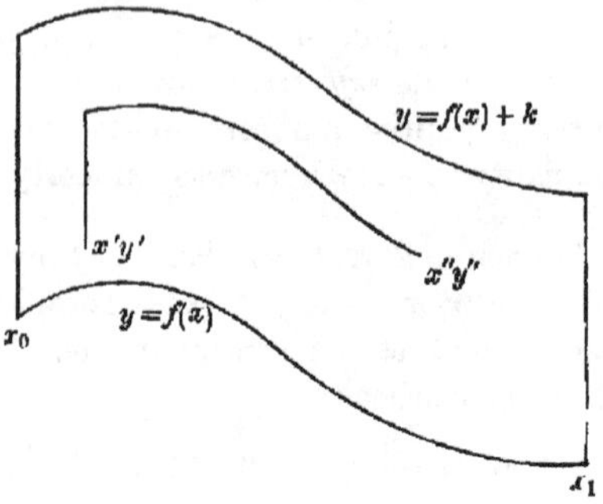

Choose $\epsilon > 0$, so that

$$2\epsilon < r' < k - 2\epsilon \quad \text{.............................(2).}$$

Since f is continuous we may choose $\delta > 0$, so that

$$|f(x) - f(x')| < \epsilon \quad \text{.............................(2a),}$$

when $|x - x'| < \delta$. It is convenient to take δ so small that

$$x_0 + \delta < x' < x_1 - \delta \quad \text{.............................(3).}$$

Then $x_0 < x < x_1$ since $|x - x'| < \delta$.

Also, when $|x - x'| < \delta$,

$$f(x) - \epsilon < f(x') < f(x) + \epsilon \quad \text{.............................(3a),}$$

so that if y be any number such that

$$y' - \epsilon < y < y' + \epsilon \quad \text{.............................(4),}$$

then
$$f(x') + r' - \epsilon < y < f(x') + r' + \epsilon \quad \text{.............................(4a).}$$

Adding (2), (3a) and (4a), we see that

$$f(x) < y < f(x) + k.$$

Therefore the point $z = x + iy$, chosen in this manner, is a point of the set S. Hence, if δ' be the smaller of δ and ϵ, and if

$$|z - z'| < \delta',$$

the conditions (2a) and (4) are both satisfied, and hence z is a member of the set. The first condition for a continuum is, consequently, satisfied.

Further, the points z', z'' (for which $r' \leq r''$), belonging to S, can be joined by the simple curve made up of the two curves defined by the relations

(i) $x = x', \quad (y' \leq y \leq y' + r'' - r')$,

(ii) $y = f(x) + r'', \quad (x' \leq x \leq x'' \text{ or } x'' \leq x \leq x')$.

Hence S is a continuum.

6. LEMMA. *Any limiting point Q of a set of points on a simple curve lies on the curve*

Take any sequence of the set which has Q as its unique limiting point; let the parameters of the points of the sequence be $t_1, t_2, \ldots$ Then the sequence (t_n) has at least[13] one limit τ, and $t_0 \leqslant \tau \leqslant T$. Since $x(t)$, $y(t)$ are continuous functions, $\lim x(t_n) = x(\tau)$, $\lim y(t_n) = y(\tau)$; and $(x(\tau), y(\tau))$ is on the curve since $t_0 \leqslant \tau \leqslant T$; i e Q is on the curve.

Corollary. If Q_0 be a fixed point not on the curve, the distance of Q_0 from points on the curve has a positive lower limit δ For if δ did not exist we could find a sequence (P_n) of points on the curve such that $Q_0P_{n+1} < \frac{1}{2} Q_0 P_n$, so that Q_0 would be a limiting point of the sequence and would therefore lie on the curve.

THEOREM I. *If a point is of order n with respect to a closed simple curve, all points sufficiently near it are of order n.*

Let Q_0 be a point not on the curve and Q_1 any other point. Then the distance of points on the curve from Q_0 has a positive lower limit, δ; so that, if $Q_0Q_1 \leqslant \frac{1}{2}\delta$, the line Q_0Q_1 cannot meet the curve

Let t be the parameter of any point, P, on the given curve, and τ the parameter of a point, Q, on Q_0Q_1, and $\theta(t, \tau)$ the angle QP makes with the axis of x; then $\theta(t, \tau)$ is a continuous function of τ, when t is fixed; therefore

$$\theta(t + \omega, \tau) - \theta(t, \tau)$$

is a continuous function[14] of τ; but the order of a point (being an integer) can only vary *per saltus*; therefore $\theta(t + \omega, \tau) - \theta(t, \tau)$ is a constant, so far as variations of τ are concerned, therefore the orders of Q_0, Q_1 are the same.

The above argument has obviously proved the following more general theorem:

THEOREM II. *If two points Q_0, Q_1 can be joined by a simple curve having no point in common with a given closed simple curve, the orders of Q_0, Q_1 with regard to the closed curve are the same.*

The following theorem is now evident:

THEOREM III. *If two points Q_0, Q_1 have* different *orders with regard to a given closed simple curve*, every *simple curve joining them has at least one point in common with the given closed curve.*

THEOREM IV. *Within an arbitrarily small distance of any point, P_0, of a regular closed curve, there are two points whose orders differ by unity.*

The curve consists of a finite number of parts, each of which can be

[13] Young, *Sets of Points*, pp. 18, 19. [14] See note 11 on p 6.

represented either by an equation of the form $y=f(x)$ or else by one of the form $x=f(y)$, where f is *single-valued* and continuous. First, let P_0 be not an end point of one of these parts.

Let the part on which P_0 lies be represented by an equation of the form $y=f(x)$; if the equation be $x=f(y)$, the proof is similar.

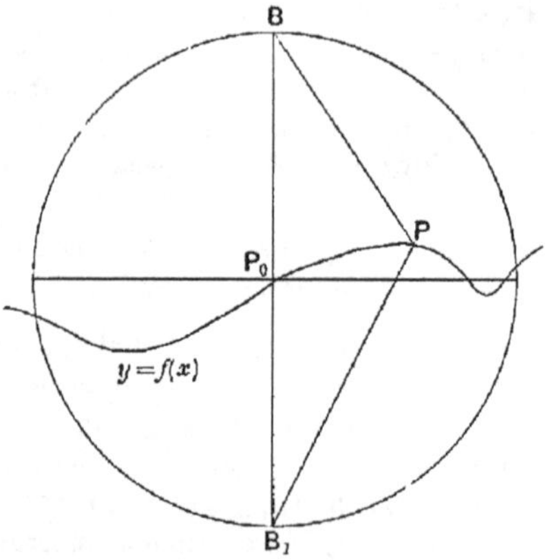

The lower limit of the distance of P_0 from any other part of the curve[15] is, say, r_1, where $r_1 > 0$.

Hence if $0 < r < r_1$, a circle of radius r, centre P_0, contains no point of the complete curve except points on the curve $y=f(x)$; and the curve $y=f(x)$ meets the ordinate of P_0 in no point except P_0.

Let B be the point (x_0, y_0+r), B_1 the point (x_0, y_0-r).

If P be any point of the curve whose parameter is t and if $\theta(t)$, $\theta_1(t)$ be the angles which BP, B_1P make with the x axis, it is easily verified that if $BP=\rho$, $B_1P=\rho_1$ and $\phi=\theta(t)-\theta_1(t)$,

$$\sin\phi = -\frac{2r(x-x_0)}{\rho\rho_1}, \quad \cos\phi = \frac{(x-x_0)^2+(y-y_0)^2-r^2}{\rho\rho_1}.$$

If ω be the period of the pair of functions $x(t)$, $y(t)$ and if δ be so small that the distances from P_0 of the points whose parameters are $t_0 \pm \delta$ are less than r, then[16], if $x(t_0+\delta) > x(t_0)$,

$$\phi(t_0) = (2n_1+1)\pi, \qquad \phi(t_0+\delta) > (2n_1+1)\pi,$$
$$\phi(t_0+\omega-\delta) < (2n_2+1)\pi, \qquad \phi(t_0+\omega) = (2n_2+1)\pi.$$

[15] If a *positive* number r_1 did not exist, by the corollary of the Lemma, P_0 would coincide with a point on the remainder of the curve; i.e. the complete curve would have a double point, and would not be a simple curve.

[16] If $x(t_0+\delta) < x(t_0)$, the inequalities involving ϕ have to be reversed.

But when $t_0 < t < t_0 + \omega$, $\sin\phi$ vanishes only when $x - x_0 = 0$, and then $\cos\phi$ is positive since $(x-x_0)^2 + (y-y_0)^2 > r^2$.

Hence $\phi \neq (2n+1)\pi$ when $t_0 < t < t_0 + \omega$; therefore since $\phi(t)$ is a continuous function of t, $n_2 - n_1 = 0$ or ± 1. But $n_1 \neq n_2$; for if $n_1 = n_2$ then $\phi(t_0+\delta) > (2n_1+1)\pi$, $\phi(t_0+\omega-\delta) < (2n_1+1)\pi$ and $\phi(t)$ would equal $(2n_1+1)\pi$ for some value of t between $t_0+\delta$ and $t_0+\omega-\delta$.

Therefore $n_2 - n_1 = \pm 1$, and consequently

$$\{\theta(t_0+\omega) - \theta(t_0)\} - \{\theta_1(t_0+\omega) - \theta_1(t_0)\} = \phi(t_0+\omega) - \phi(t_0) = \pm 2\pi,$$

that is to say the orders of B, B_1 differ by unity.

The theorem is therefore proved, except for end points of the curve.

If P_1 be an end point, a point P_0 of the curve (not an end point) can be found such that P_1P_0 is arbitrarily small; then $P_0B < P_0P_1$ since $P_0B < r_1 < P_0P_1$, and therefore $P_1B < 2P_0P_1$, so that P_1B, and similarly P_1B_1, are arbitrarily small; since the orders of B, B_1 differ by unity the theorem is proved.

THEOREM V. (i) *If two continua C_1, C_2 have a point Q in common, the set of points, S, formed by the two continua is one continuum; and* (ii) *if the two continua C_1, C_2 have no point in common, but if all points sufficiently near any point, the end points excepted, of the elementary curve $y=f(x)$, $(x_0 \leqslant x \leqslant x_1)$, belong to C_1 or to C_2, or to the curve, the points sufficiently near and above*[17] *the curve belonging to C_1 and those sufficiently near and below it to C_2, then the set of points S consisting of C_1, C_2 and the curve (the end points excepted) is one continuum.*

(i) Let P be any point of S; if P belong to, say, C_1 all points sufficiently near P belong to C_1 and therefore to S. Hence S satisfies the first condition for a continuum. Again if P, P' be any two points of S, if P, P' belong both to C_1 or both to C_2, they can be joined by a simple curve lying wholly in C_1 or C_2, i.e. wholly in S. If P belong to C_1 and P' to C_2, each can be joined to Q by a simple curve lying wholly in S. If the curves PQ, $P'Q$ have no point in common, save Q, PQP' is a simple curve lying in S. If PQ, $P'Q$ have a point in common other than Q, let PQ_1 be an arc of PQ such that Q_1 lies on $P'Q$ but no other point of PQ_1 lies on $P'Q$.

[The point Q_1 exists; for a set of points common to both curves exists; let τ be the lower boundary[18] of the parameters of the set, regarded as points on PQ; by the lemma given above, the point Q_1 with parameter τ is on both curves, and satisfies the necessary condition.]

17 The terms 'above' and 'below' are conventional: (x, y) is above (x, y') if $y > y'$.

18 The lower boundary exists. Hobson, *Functions of a Real Variable*, p. 58.

Then PQ_1, Q_1P' are simple curves with no point in common save Q_1. Hence PQ_1P' is a simple curve lying wholly in S. In either case, S satisfies the second condition for a continuum. Hence S is a continuum.

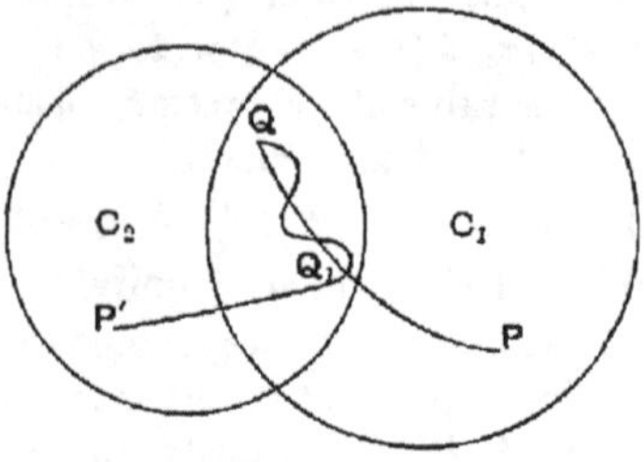

(ii) Let the curve be AB; draw CED parallel to Oy through any point E of AB (the end points excepted). If C and D be sufficiently near to AB, C belongs to C_1 and D to C_2.

Then all points sufficiently near any point of C_1 or of C_2 belong to S; and all points sufficiently near any point of AB (the end points excepted) belong to S. Hence S satisfies the first condition for a continuum.

Let P, P' belong to S. Then either[19] (*a*) P, P' both belong to C_1 or to C_2; (*b*) P belongs to C_1, P' to C_2; (*c*) P belongs to C_1, P' to AB; (*d*) P, P' both belong to AB.

In cases (*a*) and (*d*), PP' can obviously be joined by a simple curve lying wholly in S. In case (*b*), simple curves PC, CD, DP' can be drawn lying in S, and a simple curve can be drawn joining PP'. In case (*c*), simple curves PC, CE, EP' (the last being an arc of AB) can be drawn lying in S, and a simple curve can be drawn joining PP'. Hence S always satisfies the second condition for a continuum. Therefore S is a continuum.

THEOREM VI. *Given a continuum R and an elementary curve AB, then*: (*a*) *If R contain all points of the curve except possibly its end points, which may lie on the boundary of R, the set, R^-, of points of R which do not lie on AB form, at most,* two *continua.*

(*b*) *If one or both end points lie in R, R^- is* one *continuum.*

(*a*) Let the equation of AB be $y=f(x)$. Through any point of AB (not an end point) draw a line CD, parallel to Oy, bisected at the

[19] There are several other cases which are obviously equivalent to one of these; e.g. P belongs to C_2, P' to AB.

point on AB, and lying wholly in R; choose C, D so that the ordinate of C is greater than the ordinate of D.

Then R^- satisfies the first condition for a continuum (for if z_0 be a point of R^- we can choose δ so that all points satisfying $|z - z_0| < \delta$ belong to R, and since z_0 is not on AB, we can choose δ' smaller still if necessary, so that no point, z, of AB satisfies $|z - z_0| < \delta'$). Also R^- satisfies the second condition *unless a point P of R^- exists which cannot be joined to D by a simple curve lying wholly in R^-*. For if there is no such point, then if P, P' be any two points of R^-, they can each be joined to D by a simple curve; if these two curves do not intersect except at D, PDP' is a simple curve; if the two curves do intersect, let Q be the first point of intersection arrived at by a point which describes the curve PD. Then PQ, QP' are two simple curves with no point in common except Q, so that PQP' is a simple curve lying wholly in R^-; hence R^- satisfies the second condition for a continuum.

Otherwise, join P to D by a simple curve lying wholly in R; then this curve has at least one point not in R^-; i.e. it has at least one point in common with AB.

Let E be the first point on AB which is reached by a point describing the curve PD; so that PE has no point on AB except E.

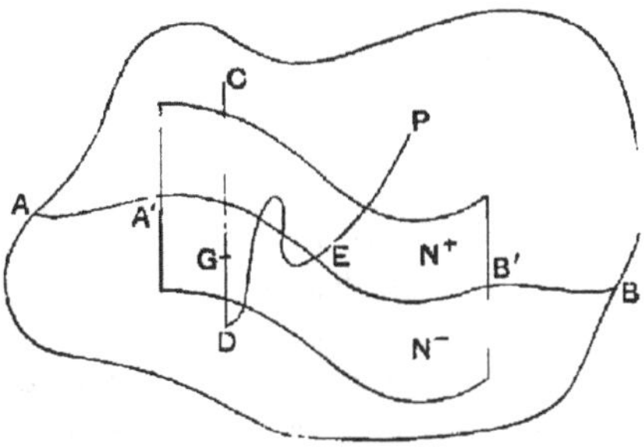

Choose an arc $A'B'$ of AB, which contains E but not A or B. Construct two continua N^+ and N^- above and below $A'B'$ respectively as in the example of § 5, each continuum lying wholly in R. Then N^+, N^- and the curve $A'B'$ with the end points omitted obviously form one continuum, so that if a point F be taken on EP sufficiently near E, it will lie on N^+ or N^-; for F cannot lie on $A'B'$. Suppose that F lies in N^-; choose a point G on CD lying in N^-; then FG can be joined by a simple curve lying in N^-. Now PF, FG, GD are three simple curves lying in N^+ and N^-; and hence a simple curve $PFGD$ can be drawn lying in N^+ or N^-; i.e. PD has been joined by a simple curve lying in R^-; but this is impossible. Hence F *must* lie in N^+: and then it can be shewn by similar reasoning that P can be joined to C by a simple curve lying wholly in R^-.

Hence the points of R^- can be divided into two sets:

(i) The points which *cannot* be joined to D by a simple curve lying wholly in R^-; these points *can* be joined to C by a simple curve lying wholly in R^-.

(ii) The points of R^- which *can* be joined to D by a simple curve lying wholly in R^-.

Each of these sets is easily seen to satisfy both the conditions for a continuum. Hence the points of R^- form *at most* two continua.

(*b*) If B lies in R, a line BB_1 may be drawn parallel to Ox lying wholly in R. Then by (*a*) the points of R not on ABB_1 form at most two continua; if they form only one continuum, the theorem is granted; for this continuum with the points on BB_1 (B excepted) forms one continuum; if they form two continua[20], these two continua with the boundary points BB_1 (B excepted) form one continuum by Theorem V.

7. THE MAIN THEOREM. *The points of the plane not on a given regular closed curve form two continua of which the entire curve is the complete boundary.*

Within an arbitrarily small distance of any point of the curve there are two points of different orders with regard to the curve, by Theorem IV of § 6. Hence by Theorem III of § 6, the points of the plane not on the curve form *at least* two continua. Divide the curve into a finite number of elementary curves and take these in the order in which they occur on the curve as t increases from t_0 to T; then by the second part of Theorem VI of § 6 each of these elementary curves, except the last, does not divide the region consisting of the plane less the points of the elementary curves already taken; the last divides the plane into *at most* two continua, by the first part of Theorem VI of § 6. Hence there are *exactly* two continua; and the points of these two continua are of different orders with regard to the curve.

[20] It is easily seen that if there are two continua the points of one of them, which are sufficiently near BB_1, are above BB_1, while the points of the other, which are sufficiently near BB_1, are below BB_1.

Any point of the curve is a boundary point of either continuum, by Theorems III and IV of § 6; and any point not on the curve is a point of one continuum by Theorem I of § 6, and is therefore not a boundary point.

8. THEOREM I. *All sufficiently distant points are of order zero with regard to a given regular closed curve.*

Let $P(x, y)$ be any point on the curve, and $P_1(x_1, y_1)$ be any other point; the angle which PP_1 makes with the axis of x is given by

$$\cos\theta = \kappa(x - x_1), \quad \sin\theta = \kappa(y - y_1),$$

where $$\kappa = \{(x - x_1)^2 + (y - y_1)^2\}^{-\frac{1}{2}}.$$

If $x_1^2 + y_1^2$ be sufficiently large, either $|x_1|$ or $|y_1|$ must be so large that either $\cos\theta$ or $\sin\theta$ never vanishes; hence the change in θ as P goes round the curve cannot be numerically so great as π; but this change is $2n\pi$ where n is an integer and is the order of P_1; hence $n = 0$.

That continuum which contains these sufficiently distant points is called the *exterior* of the curve; the other continuum is called the *interior*.

Since the order of any point of the interior of a regular closed curve differs from the order of any point of the exterior by unity, the order of any point of the interior is ± 1. If the order of any point of the interior is $+1$, we say that the point $(x(t), y(t))$ 'describes the curve in the *counterclockwise* direction as t increases from t_0 to T.'

If the order be -1, we say that the point describes the curve in the *clockwise* direction.

Let $t' = -t$; and let $\theta'(t')$ be the angle that AP makes with the axis of x, A being a point of the interior and P being the point whose parameter is t or t'.

Then $\theta'(t') - \theta(t) - 2m\pi$, and, if we take $\theta'(t')$ to vary continuously as t varies continuously, m is constant, since m can only vary *per saltus*. Consequently

$$\theta'(-T + \omega) - \theta'(-T) = -\{\theta(t_0 + \omega) - \theta(t_0)\}.$$

Therefore the order of the interior point when t' is the parameter is minus the order of the point when t is the parameter.

DEFINITION. *Oriented curves, Orientation.* Let P, S be the end points of a simple curve. Let one of them, say P, be called the *first point*. If Q, R be two other points on the curve Q is said to be *before* R if $t_P < t_Q < t_R$ or if $t_P > t_Q > t_R$. The points of the curve have thus been *ordered*[21]. Such an ordered set of points PS is called an *oriented curve*; it differs from the oriented curve SP in which S is the first point.

Two oriented curves C_1, C_2 with a common arc σ have the *same orientation* if the points of σ are in the same order whether σ is regarded as belonging to C_1 or to C_2. If the points are not in the same order, the curves have *opposite orientations*.

It is easy to see that if P, Q, R be three points on a regular closed curve, the curves $PQRP$, $PRQP$ have opposite orientations.

We agree to choose the parameter of an oriented curve so that the first point has the smallest parameter. This can be done by taking a new parameter $t' = -t$, if necessary.

It is convenient *always* to choose that orientation of a closed curve which makes the order of interior points $+1$; that is to say that an oriented closed curve is such that a point describes it counterclockwise as t increases from t_0 to $t_0 + \omega$.

THEOREM II[22]. *Let two continua R_1, R_2 be the interiors of two regular closed curves C_1, C_2 respectively. Let a segment σ_1 of C_1 coincide with a segment σ_2 of C_2; then* (i) *if R_1, R_2 have no point in common the orientations of σ_1, σ_2 are opposite*; *and* (ii) *if R_1 be wholly interior*[23] *to R_2, the orientations of σ_1, σ_2 are the same.*

(i) If the orientations of σ_1 and σ_2 are the same, by Theorem IV of § 6 it follows that arbitrarily near any point P_0 of σ_1 and σ_2 (not an end point) there are two points B, B' such that the order of B with regard to either C_1 or C_2 exceeds that of B' by unity, so that B is an interior point of both curves which is impossible. Hence the orientation of σ_1 is opposite to that of σ_2.

(ii) If the orientations of σ_1, σ_2 are different, we can find points B, B' arbitrarily near any point P_0 of σ_1 and σ_2 such that (*a*) the order of B' with regard to C_1 exceeds that of B by $+1$, (*b*) the order of B with regard to C_2 exceeds that of B' by $+1$. Consequently B is a point of R_1 but not of R_2; this is impossible. Hence the orientations of σ_1, σ_2 are the same.

[21] Hobson, *Functions of a Real Variable*, § 122.

[22] Ames points out that Goursat tacitly assumes this theorem.

[23] I e. if every point of R_1 is a point of R_2.

CHAPTER II

COMPLEX INTEGRATION

§ 9. The integral of a function of a real variable; extension to complex variables; restriction of the path of integration.—§ 10. Definition of a complex integral.—§ 11. Existence theorems.—§ 12. Goursat's lemma.—§ 13. Various simple theorems.

9. The integral[1] of a continuous function, $f(x)$, of a real variable x, is defined by means of the limit of a sum in the following manner:

Divide an interval a to b ($a \leqslant b$) into 2^n equal parts and let γ_r be the rth part. Let H_r, h_r be the upper and lower limits of $f(x)$ in γ_r; let

$$S_n = \sum_{r=1}^{n} H_r \cdot \frac{b-a}{2^n}, \quad s_n = \sum_{r=1}^{n} h_r \cdot \frac{b-a}{2^n}.$$

Then (S_n) is a non-increasing sequence and (s_n) is a non-decreasing sequence, and $S_n \geqslant s_n$, consequently S_n, s_n have finite limits as $n \to \infty$; and if $f(x)$ is continuous it can be proved that these two limits are the same; the common value of these two limits is called the integral of $f(x)$ taken between the end-values or limits a and b, and is written

$$\int_a^b f(x)\,dx.$$

Further, it can be shewn that if ϵ is arbitrary, a number δ can be found such that if the interval a to b be divided into *any* sub-intervals $\eta_1, \eta_2, \ldots \eta_\nu$ each less than δ, and if x_r be *any* point in the rth interval, then

$$\left| \int_a^b f(x)\,dx - \sum_{r=1}^{\nu} \eta_r f(x_r) \right| < \epsilon.$$

When we study the theory of functions of complex variables, we naturally enquire whether it is not possible to generalise this definition; for the interval a to b may be regarded as a segment of a particular curve in the Argand diagram, namely the real axis.

[1] Bromwich's *Theory of Infinite Series* (1908), §§ 157–163, should be consulted; the analysis given above is quoted from § 163.

This suggests that we should define the integral of a continuous function, $f(z)$, of the complex variable z, taken along a curvilinear path AB in the Argand diagram by the natural extension of the above definition, namely that the integral of $f(z)$, taken between the limits z_0 and Z, is the number S (if that number exist) such that it is possible to make

$$\left| S - \sum_{r=0}^{\nu} (z_{r+1} - z_r) f(z_r') \right|$$

less than an arbitrary positive number ϵ by taking ν points $z_1, z_2, \ldots z_\nu$ in order on the curve AB ($z_{\nu+1}$ being interpreted as meaning Z) in any way such that

$$| z_{r+1} - z_r | < \delta \text{ for } r = 0, 1, 2, \ldots \nu,$$

δ being a number depending on ϵ (so that ν also depends on ϵ), and the point z_r' being any point on the curve between z_r and z_{r+1}.

[Note that we do not say

$$S = \lim_{\nu \to \infty} \sum_{r=0}^{\nu} (z_{r+1} - z_r) f(z_r'),$$

because the summation on the right is a function of $2\nu + 1$ independent variables $z_1, z_2, \ldots z_\nu, z_0', z_1', \ldots z_\nu'$, and so S is not an ordinary limit of a function of one variable.]

It is, however, necessary to define exactly what is meant by the phrase 'points in order on the curve AB.'

To ensure that the limit, by which we shall define an integral, may exist, we shall restrict the curve on which the points $z_1, z_2, \ldots$ lie, to be an 'oriented simple curve.' And a further restriction is convenient, namely that the curve should have limited variations[2]; that is to say that the functions $x(t)$, $y(t)$ should have limited variations in the interval t_0 to T.

[It can be proved[3] that a necessary and sufficient condition that a simple curve should have a finite length is that it should have limited variations, but this proposition will not be required; the lemma below will be sufficient for the purposes of this work.]

A function $f(z)$ of a complex variable z is said to be 'continuous on a simple curve' if $f(z)$ is a continuous function of t.

[2] Young, *Sets of Points*, §§ 140–141. Jordan, *Cours d'Analyse*, t. I. p. 90. It will be obvious that the definition may be extended to cover the case when the path of integration consists of a finite number of simple curves with limited variations.

[3] Young, *Sets of Points*, § 167.

We can now prove the following important lemma:

LEMMA. *Let $z_0, z_1, z_2, \ldots z_{n+1}$ be any sequence of points in order on a simple curve. Then* $\sum_{r=0}^{n} |(z_{r+1} - z_r)|$ *is less than or equal to the sum of the total variations of $x(t)$ and $y(t)$ as t varies from t_0 to t_{n+1}.*

Since the modulus of a sum does not exceed the sum of the moduli, it follows that

$$\sum_{r=0}^{n} |(z_{r+1} - z_r)| = \sum_{r=0}^{n} |\{(x_{r+1} - x_r) + i(y_{r+1} - y_r)\}|$$

$$\leqslant \sum_{r=0}^{n} [|(x_{r+1} - x_r)| + |\{i(y_{r+1} - y_r)\}|]$$

$$\leqslant \sum_{r=0}^{n} |\{x(t_{r+1}) - x(t_r)\}| + \sum_{r=0}^{n} |\{y(t_{r+1}) - y(t_r)\}|.$$

But $t_{r+1} \geqslant t_r$, since the points $z_0, z_1, z_2, \ldots$ are in order; and consequently the first of these summations is less than or equal to the total variation of $x(t)$, and the second summation is less than or equal to the total variation of $y(t)$; that is to say, $\Sigma|(z_{r+1} - z_r)|$ is less than or equal to the sum of the total variations of $x(t)$ and $y(t)$.

10. We are now in a position to give a formal definition of a complex integral and to discuss its properties. The notation which has been introduced in §§ 2, 4 and 5 will be employed throughout.

DEFINITION. *Let AB be a simple curve with limited variations drawn in the Argand diagram. Let $f(z)$ be a function of the complex variable z which is continuous on the curve AB. Let z_0 be the complex coordinate of A, and Z the complex coordinate of B. Let a sequence of points on AB be chosen, and when n of these points have been taken, let the points* taken in order *be called $z_1^{(n)}, z_2^{(n)}, \ldots z_n^{(n)}$ (so that if $m \geqslant n$, $z_1^{(n)}$ is one of the points $z_1^{(m)}, z_2^{(m)}, \ldots z^{(n)}_{m-n+1}$); the sequence of points may be chosen according to any definite law whatever*[4], *provided only that the points are all different and that, given any positive number δ, we can find an integer n_0 such that when $n \geqslant n_0$,*

$$0 < t_{r+1}^{(n)} - t_r^{(n)} \leqslant \delta,$$

where $r = 0, 1, 2, \ldots n$ and $t_0^{(n)} = t_0$, $t_{n+1}^{(n)} = T$.

[4] If $t_0 = 0$, $T = 1$, the simplest law is given by taking $t_1^{(n)}, t_2^{(n)}, \ldots t_n^{(n)}$ to be the first n of the numbers $\frac{1}{2}$; $\frac{1}{4}$, $\frac{3}{4}$; $\frac{1}{8}$, $\frac{3}{8}$, $\frac{5}{8}$, $\frac{7}{8}$; when these n numbers are rearranged in order of magnitude.

Then the complex integral $\int_A^B f(z)\,dz$ *is defined as meaning the following limit:*

$$\int_A^B f(z)\,dz = \lim_{n\to\infty} [(z_1^{(n)} - z_0^{(n)}) f(z_0^{(n)}) + (z_2^{(n)} - z_1^{(n)}) f(z_1^{(n)}) + (z_3^{(n)} - z_2^{(n)}) f(z_2^{(n)}) + \ldots + (Z - z_n^{(n)}) f(z_n^{(n)})]$$

$$= \lim_{n\to\infty} \sum_{r=0}^{n} [(z_{r+1}^{(n)} - z_r^{(n)}) f(z_r^{(n)})].$$

[It is permissible to speak of the limit of

$$\sum_{r=0}^{n} [(z_{r+1}^{(n)} - z_r^{(n)}) f(z_r^{(n)})]$$

because these expressions form a sequence (depending on n), each member of the sequence being determinate when the form of f and the law, by which the points $z_r^{(n)}$ are chosen, are given.]

The integral is said to be taken along the path AB, and the path AB is usually called the contour of integration; and if the path AB be called C, we sometimes write $\int_A^B f(z)\,dz$ in the form $\int_{(AB)} f(z)\,dz$ or $\int_C f(z)\,dz$.

11. It is next necessary to prove (Theorem I) that the limit, by which an integral is defined, exists.

When we have proved Theorem I we shall prove (Theorem II) that if a positive number ϵ be taken arbitrarily, it is possible to find a number δ_1 such that, when *any* ν numbers $t_1, t_2, \ldots t_\nu$ are taken so that $t_0 \leqslant t_1 \leqslant t_2 \leqslant \ldots \leqslant t_\nu \leqslant t_{\nu+1} = T$ and $t_{p+1} - t_p \leqslant \delta_1$ $(p = 0, 1, \ldots \nu)$, and when T_p is such that $t_p \leqslant T_p \leqslant t_{p+1}$, then

$$\left| \int_A^B f(z)\,dz - \sum_{p=0}^{\nu} (z_{p+1} - z_p) f(Z_p) \right| < \epsilon.$$

THEOREM I. *Let* $S_n(z) = \sum_{r=0}^{n} [(z_{r+1}^{(n)} - z_r^{(n)}) f(z_r^{(n)})]$; *then* $\lim_{n\to\infty} S_n(z)$ *exists.*

To prove the existence of the limit, we shall prove that, given an arbitrary positive number ϵ, we can choose an integer n such that, when $m > n$,

$$| S_m(z) - S_n(z) | < \epsilon.$$

This establishes[5] the existence of $\lim_{n\to\infty} S_n(z)$.

[5] Bromwich, *Theory of Infinite Series*, §§ 3, 75, 151.

Let L be the sum of the total variations of $x(t)$ and $y(t)$ for the interval t_0 to T of t.

In virtue of the continuity of $f(z)$ *qua* function of t, corresponding to an arbitrary positive number ϵ, we can find a positive number δ such that, if z be any particular point on AB, and if z' be on AB, then[6]

$$|f(z') - f(z)| \leqslant \tfrac{1}{4}\epsilon/L \quad \ldots\ldots\ldots\ldots (5)$$

whenever $|t' - t| \leqslant \delta$; it is obvious that, in general, δ is a function of t.

Let us *assume* for the present[7] that, when ϵ is taken arbitrarily, a number δ_0 (independent of t, but depending on ϵ) exists, such that, for *all* values of t under consideration,

$$\delta \geqslant \delta_0 > 0;$$

that is to say, we assume that $f(z)$ is a *uniformly continuous*[8] function of t.

Now choose n so large that

$$0 < t_{r+1}^{(n)} - t_r^{(n)} \leqslant \delta_0,$$

for $r = 0, 1, 2, \ldots n$; this is possible by reason of the hypothesis made concerning the law by which the numbers $t_r^{(n)}$ were chosen.

Let m be any integer such that $m > n$; and let those of the points $z_s^{(m)}$ which lie between $z_0^{(n)}$ and $z_1^{(n)}$ be called $z_{1,0}, z_{2,0}, \ldots z_{m_0+1,0}$, where $z_{1,0} = z_0^{(n)}$, $z_{m_0+1,0} = z_1^{(n)}$; and, generally, let those of the points $z_s^{(m)}$ which lie between $z_r^{(n)}$ and $z_{r+1}^{(n)}$ be called $z_{1,r}, z_{2,r}, \ldots z_{m_r+1,r}$, where $z_{1,r} = z_r^{(n)}$, $z_{m_r+1,r} = z_{r+1}^{(n)}$.

Then
$$S_n = \sum_{r=0}^{n} [(z_{r+1}^{(n)} - z_r^{(n)}) \,.\, f(z_r^{(n)})]$$
$$= \sum_{r=0}^{n} \left[\left\{\sum_{s=1}^{m_r} (z_{s+1,r} - z_{s,r})\right\} f(z_r^{(n)})\right],$$

since the points $z_{s,r}$ are the same as the points $z_r^{(m)}$.

Also
$$S_m = \sum_{r=0}^{n} \left[\sum_{s=1}^{m_r} \{(z_{s+1,r} - z_{s,r}) \,.\, f(z_{s,r})\}\right],$$

so that
$$|(S_n - S_m)| = \left|\sum_{r=0}^{n} \left[\sum_{s=1}^{m_r} (z_{s+1,r} - z_{s,r})\{f(z_r^{(n)}) - f(z_{s,r})\}\right]\right|$$
$$\leqslant \sum_{r=0}^{n} \sum_{s=1}^{m_r} |(z_{s+1,r} - z_{s,r})\{f(z_r^{(n)}) - f(z_{s,r})\}|.$$

[6] The reason for choosing the multiplier $\frac{1}{4}$ will be seen when we come to Theorem II.

[7] A formal proof is given in § 12.

[8] The continuity is said to be uniform because, as $t' \to t$, $f(z')$ tends to the limit $f(z)$ uniformly with respect to the variable t.

But $t_{r+1}^{(n)} \geqslant t_{s,r} \geqslant t_r^{(n)}$, so that $0 \leqslant t_{s,r} - t_r^{(n)} \leqslant \delta_0$, and consequently $|f(z_r^{(n)}) - f(z_{s,r})| \leqslant \frac{1}{4}\epsilon/L$; also $\sum_{r=0}^{n} \sum_{s=1}^{m_r} |z_{s+1,r} - z_{s,r}| = \sum_{r=0}^{m} |z_{r+1}^{(m)} - z_r^{(m)}|$, and consequently

$$|S_n - S_m| \leqslant \sum_{r=0}^{n} \sum_{s=1}^{m_r} |(z_{s+1,r} - z_{s,r}) \tfrac{1}{4}\epsilon L^{-1}|$$

$$\leqslant \tfrac{1}{4}\epsilon L^{-1} \sum_{r=0}^{m} |(z_{r+1}^{(m)} - z_r^{(m)})|$$

$$\leqslant \tfrac{1}{4}\epsilon,$$

since

$$\sum_{r=0}^{m} |(z_{r+1}^{(m)} - z_r^{(m)})| \leqslant L,$$

by the Lemma of § 9. That is to say that, given an arbitrary positive number ϵ, we have found n such that when $m > n$, $|S_n - S_m| < \epsilon$, and consequently we have proved that $\lim_{n\to\infty} S_n$ exists; the value of this limit is written

$$\int_{z_0}^{Z} f(z)\,dz.$$

We can now prove the following general theorem:

THEOREM II. *Given any positive number ϵ, it is possible to find a positive number δ_1 such that, when any ν numbers $t_1, t_2, \ldots t_\nu$ are taken so that $0 \leqslant t_{p+1} - t_p \leqslant \delta_1$, $(p = 0, 1, \ldots \nu$, and $t_{\nu+1} = T)$, while T_p is such that $t_p \leqslant T_p \leqslant t_{p+1}$, then*

$$\left| \int_{z_0}^{Z} f(z)\,dz - \sum_{p=0}^{\nu} (z_{p+1} - z_p) f(Z_p) \right| < \epsilon,$$

z_p, Z_p being the points whose parameters are t_p, T_p respectively.

Choose δ_0 and n to depend on ϵ in the same way as in the proof of Theorem I; we shall prove that *it is permissible to take $\delta_1 = \delta_0$.*

For, assuming that $0 \leqslant t_{p+1} - t_p \leqslant \delta_0$, we can find an integer r corresponding to each of the numbers t_p, $(p \neq \nu + 1)$, such that $t_r^{(n)} \leqslant t_p < t_{r+1}^{(n)}$; let the numbers t_p which satisfy this inequality for any particular value of r be called in order $t_{1,r}, t_{2,r}, \ldots t_{N_r,r}$.

Then we may write

$$\sum_{p=0}^{\nu} [(z_{p+1} - z_p) f(Z_p)] = \sum_{r=0}^{n} [(z_{1,r} - z_r^{(n)}) f(Z_{0,r}) + (z_{2,r} - z_{1,r}) f(Z_{1,r})$$
$$+ (z_{3,r} - z_{2,r}) f(Z_{2,r}) + \ldots + (z_{r+1}^{(n)} - z_{N_r,r}) f(Z_{N_r,r})].$$

The following conventions have to be adopted in interpreting the summation on the right-hand side:

(i) $t_{0,r} \leqslant T_{0,r} \leqslant t_{1,r}$; where $t_{0,r}$ means that number of the set $t_0, t_1, \ldots t_r$ which immediately precedes $t_{1,r}$.

(ii) $t_{N_r,r} \leqslant T_{N_r,r} \leqslant t_{N_r+1,r}$; where $t_{N_r+1,r}$ means that number of the set $t_1, t_2, \ldots t_{r+1}$ which immediately follows $t_{N_r,r}$.

(iii) If, for any value of r, there is no number t_p such that $t_r^{(n)} \leqslant t_p < t_{r+1}^{(n)}$, the term of the summation corresponding to that value of r is $(z_{r+1}^{(n)} - z_r^{(n)}) f(Z_{0,r})$, where $t_{0,r} \leqslant T_{0,r} \leqslant t_{1,r}$ and $t_{0,r}$, $t_{1,r}$ are respectively the largest and smallest numbers of the set $t_0, t_1, \ldots t_{\nu+1}$ which satisfy the inequalities

$$t_{0,r} < t_r^{(n)}, \quad t_{1,r} \geqslant t_{r+1}^{(n)}.$$

With these conventions, if S_n has the same meaning as in Theorem I, we may write

$$\sum_{p=0}^{\nu} (z_{p+} - z_p) f(Z_p) - S_n$$

$$= \sum_{r=0}^{n} [(z_{1,r} - z_r^{(n)})\{f(Z_{0,r}) - f(z_r^{(n)})\} + (z_{2,r} - z_{1,r})\{f(Z_{1,r}) - f(z_r^{(n)})\}$$
$$+ \ldots + (z_{r+1}^{(n)} - z_{N_r,r})\{f(Z_{N_r,r}) - f(z_r^{(n)})\}];$$

if for any value of r, there is no number t_p such that $t_r^{(n)} \leqslant t_p < t_{r+1}^{(n)}$, the term of the summation corresponding to that value of r is

$$(z_{r+1}^{(n)} - z_r^{(n)})\{f(Z_{0,r}) - f(z_r^{(n)})\}.$$

Now if $s = 0, 1, \ldots N_r$, we have

$$T_{s,r} \leqslant t_{s,r} + \delta_0 < t_{r+1}^{(n)} + \delta_0 < t_r^{(n)} + 2\delta_0,$$

and $$T_{s,r} \geqslant t_{s+1,r} - \delta_0 \geqslant t_r^{(n)} - \delta_0;$$

hence $$|T_{s,r} - t_r^{(n)}| < 2\delta_0.$$

Therefore, if $t' = \frac{1}{2}(T_{s,r} + t_r^{(n)})$, we have

$$|T_{s,r} - t'| < \delta_0, \qquad |t' - t_r^{(n)}| < \delta_0;$$

so that, since the modulus of a sum does not exceed the sum of the moduli,

$$|f(Z_{s,r}) - f(z_r^{(n)})| \leqslant |f(Z_{s,r}) - f(z')| + |f(z') - f(z_r^{(n)})|$$
$$\leqslant \tfrac{1}{2}\epsilon L^{-1},$$

by equation (5) of Theorem I.

It follows that

$$\left|\sum_{p=0}^{\nu}(z_{p+1}-z_p)f(Z_p)-S_n\right| \leqslant \sum_{r=0}^{n}\left[\left|(z_{1,r}-z_r^{(n)})\right|.\tfrac{1}{2}\epsilon L^{-1}\right.$$

$$\left.+\left|(z_{2,r}-z_{1,r})\right|.\tfrac{1}{2}\epsilon L^{-1}+\ldots+\left|(z_{r+1}^{(n)}-z_{N_r,r})\right|.\tfrac{1}{2}\epsilon L^{-1}\right]$$

$$\leqslant \tfrac{1}{2}\epsilon L^{-1}\sum_{r=0}^{n}\left[\left|(z_{1,r}-z_r^{(n)})\right|+\left|(z_{2,r}-z_{1,r})\right|+\ldots+\left|(z_{r+1}^{(n)}-z_{N_r,r})\right|\right].$$

Now, by the Lemma of § 9, the general term of this last summation is less than or equal to the sum of the variations of $x(t)$ and $y(t)$ in the interval $t_r^{(n)}$ to $t_{r+1}^{(n)}$, since the points

$$z_r^{(n)},\; z_{1,r},\; z_{2,r},\; \ldots\; z_{N_r,r},\; z_{r+1}^{(n)}$$

are in order; and, hence, since the numbers $t_0^{(n)}, t_1^{(n)}, \ldots t_{n+1}^{(n)}$ are in order, the whole summation is less than or equal to the sum of the variations of $x(t)$ and $y(t)$ in the interval $t_0^{(n)}$ to $t_{n+1}^{(n)}$; that is to say

$$\left|\sum_{p=0}^{\nu}(z_{p+1}-z_p)f(Z_p)-S_n\right| \leqslant \tfrac{1}{2}\epsilon L^{-1}\times L$$

$$\leqslant \tfrac{1}{2}\epsilon.$$

But, by Theorem I, with the choice of n which has been made

$$|S_m-S_n| \leqslant \tfrac{1}{4}\epsilon,$$

when $m > n$. Hence, since ϵ is *independent* of m,

$$|(\lim_{m\to\infty} S_m)-S_n| \leqslant \tfrac{1}{4}\epsilon,$$

i.e.

$$\left|\int_{z_0}^{Z} f(z)\,dz-S_n\right| \leqslant \tfrac{1}{4}\epsilon.$$

Therefore

$$\left|\int_{z_0}^{Z} f(z)\,dz-\sum_{p=0}^{\nu}(z_{p+1}-z_p)f(Z_p)\right|$$

$$\leqslant\left|\int_{z_0}^{Z} f(z)\,dz-S_n\right|+\left|S_n-\sum_{p=0}^{\nu}(z_{p+1}-z_p)f(Z_p)\right|$$

$$\leqslant \tfrac{3}{4}\epsilon.$$

That is to say that, corresponding to an arbitrary positive number ϵ, we have been able to find a positive number δ_1 (namely, the number denoted by δ_0 in Theorem I), such that if

$$0 \leqslant t_{p+1}-t_p \leqslant \delta_1,\quad (p=0, 1, 2, \ldots \nu, \text{ and } t_{\nu+1}=T),$$

then

$$\left|\int_{z_0}^{Z} f(z)\,dz-\sum_{p=0}^{\nu}(z_{p+1}-z_p)f(Z_p)\right| < \epsilon.$$

From this general theorem, we can deduce the following particular theorem :

THEOREM III. *The value of* $\int_{z_0}^{Z} f(z)\,dz$ *does not depend on the particular law by which the points* $z_r^{(n)}$ *are chosen, provided that the law satisfies the conditions of* § 10.

Let points chosen according to any other law than that already considered be called $\zeta_p^{(\nu)}$, $(p=0, 1, \ldots \nu;\ \zeta_0^{(\nu)}=z_0,\ \zeta_{\nu+1}^{(\nu)}=Z)$; then if τ be the parameter of the point ζ, we can find a number ν_0 such that when $\nu>\nu_0$, $0\leqslant \tau_{p+1}^{(\nu)}-\tau_p^{(\nu)}\leqslant\delta_0$; hence we may take the numbers t_p of Theorem II to be the numbers $\tau_p^{(\nu)}$ respectively, and we will take $Z_p=\zeta_p^{(\nu)}$; therefore, by the result of Theorem II,

$$\left|\int_{z_0}^{Z} f(z)\,dz-\sum_{p=0}^{\nu}\left[(\zeta_{p+1}^{(\nu)}-\zeta_p^{(\nu)})\,f(\zeta_p^{(\nu)})\right]\right|<\epsilon;$$

and, corresponding to any positive number ϵ, we can always find the number ν_0 such that this inequality is satisfied when $\nu>\nu_0$.

Therefore $$\lim_{\nu\to\infty}\sum_{p=0}^{\nu}\left[(\zeta_{p+1}^{(\nu)}-\zeta_p^{(\nu)})\,f(\zeta_p^{(\nu)})\right]$$

exists[9] and is equal to $\int_{z_0}^{Z} f(z)\,dz$, which has been proved to be the value of

$$\lim_{n\to\infty}\sum_{p=0}^{n}(z_{p+1}^{(n)}-z_p^{(n)})\,f(z_p^{(n)});$$

and this is the result which had to be proved, namely to shew that the value of $\int_{z_0}^{Z} f(z)\,dz$ does not depend on the particular law by which we choose the points $z_r^{(n)}$.

12. It was assumed in the course of proving Theorem I of § 11 that if a function of a real variable was continuous at all points of a finite closed interval, then the function was uniformly continuous in the interval.

A formal proof of this assumption is now necessary[10]; but it is expedient first to prove the following Lemma. The lemma is proved for a two-dimensional region, as that form of it will be required later.

[9] Bromwich, *Theory of Infinite Series*, § 1.

[10] It was pointed out by Heine, *Crelle's Journal*, vol. LXXI (1870), p. 361 and vol. LXXIV (1872), p. 188, that it is not obvious that continuity implies *uniform* continuity.

Goursat's Lemma[11] *Given* (i) *a function of position of two points P', P, which will be written $\{P', P\}$, and* (ii) *an arbitrary positive number ϵ; let a finite two-dimensional closed region*[12] *R have the property that for each point P of R we can choose a* positive *number δ (depending on the position of P), such that $|\{P', P\}| \leqslant \epsilon$ whenever the distance PP' is less than or equal to δ, and the point P' belongs to the region.*

Then the region, R, can be divided into a finite *number of closed sets of points such that* each *set contains at least one point P_1 such that the condition $|\{P', P_1\}| \leqslant \epsilon$ is satisfied for* all *points P' of the set under consideration.*

If a set of points is such that for any particular positive number ϵ, a point P_1 can be found such that

$$|\{P', P_1\}| < \epsilon$$

for all points P' of the set, we shall say that the set satisfies condition (A). A set of points which satisfies condition (A) will be called a *suitable* set.

Let R^- be the continuum formed by the interior of R; take any point of R^- and draw a square, with this point as centre, whose sides are parallel to the axes, the lengths of the sides of the square being $2L$, where L is so large that no point of R lies outside the square.

If every point of R satisfies condition (A), what is required is proved. If not, divide the square into four equal squares by two lines through its centre, one parallel to each axis. Let the sets of points of R which lie either inside these squares or on their boundaries be called α_1, α_2, α_3, α_4 respectively of which α_1, α_2 are above α_3, α_4 and α_1, α_3 are on the left of α_2, α_4.

If these sets, α_1, α_2, α_3, α_4, each satisfy condition (A), what is required is proved. If any one of the sets, say α_1, does not satisfy condition (A), divide the square[13] of which α_1 forms part into four equal squares by lines parallel to the axes; let the sets of points of R which lie inside these squares or on their boundaries be called β_1, β_2, β_3 (in the figure one of the squares into which α_1 is divided contains no point of R).

If condition (A) is satisfied by each of the sets, we have divided α_1 into sets for which condition (A) is satisfied; if the condition (A) is not satisfied by any one of the sets, say β_3, we draw lines dividing the square (of side $\frac{1}{2}L$), of which β_3 forms part, into four equal squares of side $\frac{1}{4}L$.

This process of subdividing squares will either terminate or it will not; if it does terminate, R has been divided into a finite number of closed sets of points each satisfying condition (A), *and the lemma is proved*.

Suppose that the process does not terminate.

A closed set of points R' for which the process does terminate will be said to satisfy condition (B).

Then the set R does not satisfy condition (B), therefore at least one of the

[11] This form of the statement of Goursat's Lemma is due to Dr Baker.

[12] Consisting of a continuum and its boundary.

[13] A square which does satisfy condition (A) is not to be divided; for some of the subdivisions might not satisfy condition (A).

sets a_1, a_2, a_3, a_4 does not satisfy condition (B). Take the *first*[14] of them which does not.

The process of dividing the square, in which this set lies, into four equal parts gives at most four sets of points, of which at least one set does not satisfy condition (B). Take the first of them which does not, and continue this process of division and selection. The result of the process is to give an unending sequence of squares satisfying the following conditions:

If the sequence be called $s_0, s_1, s_2, \ldots$, then[15]

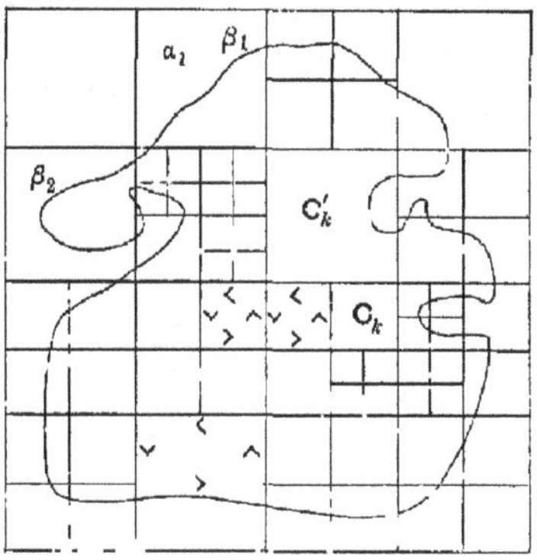

(i) The side of s_n is of length $2^{-n}L$.

(ii) No point of s_{n+1} lies outside s_n.

(iii) Two sides of s_{n+1} lie along two sides of s_n.

(iv) s_n contains at least one point of R.

(v) The set of points of R which are inside or on s_n do not satisfy condition (A).

Let the coordinates of the corners of s_n be called

$$(x_n^{(1)}, y_n^{(1)}), \quad (x_n^{(1)}, y_n^{(2)}), \quad (x_n^{(2)}, y_n^{(1)}), \quad (x_n^{(2)}, y_n^{(2)}),$$

where $$x_n^{(1)} < x_n^{(2)}, \quad y_n^{(1)} < y_n^{(2)}.$$

Then $(x_n^{(1)})$ is a non-decreasing sequence and $(x_n^{(2)})$ is a non-increasing sequence; and $x_n^{(2)} - x_n^{(1)} = 2^{-n}L$; therefore the sequences $(x_n^{(1)})$, $(x_n^{(2)})$ have a common limit ξ such that $x_n^{(1)} \leqslant \xi \leqslant x_n^{(2)}$; similarly the sequences $(y_n^{(1)})$, $(y_n^{(2)})$ have a common limit η such that $y_n^{(1)} \leqslant \eta \leqslant y_n^{(2)}$.

Consequently (ξ, η) lies inside or on the boundaries of all the squares of the sequence (s_n); further, (ξ, η) lies inside or on the boundary of the region R;

[14] We take the *first* possible square of each group of four so as to get a *definite* sequence of squares.

[15] Cp. Bromwich, *Theory of Infinite Series*, § 150.

for since s_n contains at least one point of R, the distance of (ξ, η) from at least one point of R is less than or equal to the diagonal of s_n, i.e. $2^{-n}L\sqrt{2}$. Hence, corresponding to each square, s_n, there is a point P_n such that

$$\Pi P_n \leqslant 2^{-n}L\sqrt{2},$$

where Π is the point whose coordinates are (ξ, η); this sequence of points (P_n) obviously has Π for its limiting point; and since the region R is closed, the limiting point of any sequence of points of R is a point of R. Therefore Π is a point of R.

Then $|\{P', \Pi\}| < \epsilon$ when P' is a point of R such that $P'\Pi < \delta_\pi$, where δ_π is a positive number depending on Π.

Choose n so that $2^{-n}L\sqrt{2} < \delta_\pi$; then all points, P', of s_n are such that $P'\Pi < \delta_\pi$; and therefore s_n satisfies condition (A); which is contrary to condition (v).

Consequently, by assuming that the process of dividing squares does not terminate, we are led to a contradiction; therefore all the sequences terminate; and consequently the number of sets of points into which R has to be divided is finite, that is to say, the lemma is proved.

[The reader can at once extend this lemma to space of n dimensions.]

In the one-dimensional case, the lemma is that if, given an arbitrary positive number ϵ, for each point P of a closed interval we can choose δ (depending on P) such that $|\{P', P\}| < \epsilon$ when $PP' < \delta$, then the interval can be divided into a *finite* number of sub-intervals such that a point P_1 of any sub-interval can be found such that $|\{P, P_1\}| < \epsilon$ for all points P of that sub-interval; the proof is obtained in a slightly simpler manner than in the two-dimensional case, by bisecting the interval and continually bisecting any sub-interval for which the condition (A) is not satisfied.

The proof that a continuous function of a real variable is uniformly continuous is immediate. Let $f(x)$ be continuous when $a \leqslant x \leqslant b$; we shall prove that, given ϵ, we can find δ_0 such that, if x', x'' be any two points of the interval satisfying $|x' - x''| < \delta_0$, then $|f(x') - f(x'')| < \epsilon$.

For, given an arbitrary positive number ϵ, since $f(x)$ is continuous, corresponding to any x we can find δ such that

$$|f(x') - f(x)| < \tfrac{1}{4}\epsilon \text{ when } |x' - x| < \delta.$$

Then, by the lemma, we can divide the interval a to b into a finite number of closed sub-intervals such that in each sub-interval there is a point, x_1, such that $|f(x') - f(x_1)| < \frac{1}{4}\epsilon$ when x' lies in the interval in which x_1 lies.

Let δ_0 be the length of the smallest of these sub-intervals; and let x', x'' be *any* two points of the interval $a \leqslant x \leqslant b$ such that

$$|x' - x''| < \delta_0;$$

then x', x'' lie in the same or in adjacent sub-intervals; if x', x'' lie in the same sub-interval, then we can find x_1 so that

$$|f(x')-f(x_1)|<\tfrac{1}{4}\epsilon, \quad |f(x'')-f(x_1)|<\tfrac{1}{4}\epsilon.$$

Hence
$$|f(x')-f(x'')|<\tfrac{1}{2}\epsilon.$$

If x', x'' lie in adjacent sub-intervals let ξ be their common end-point; then we can find a point x_1 in the first sub-interval and a point x_2 in the second such that

$$|f(x')-f(x_1)|<\tfrac{1}{4}\epsilon, \quad |f(\xi)-f(x_1)|<\tfrac{1}{4}\epsilon,$$
$$|f(x'')-f(x_2)|<\tfrac{1}{4}\epsilon, \quad |f(\xi)-f(x_2)|<\tfrac{1}{4}\epsilon,$$

so that

$$\begin{aligned}|f(x')-f(x'')| &= |\{f(x')-f(x_1)\}-\{f(\xi)-f(x_1)\} \\ &\qquad -\{f(x'')-f(x_2)\}+\{f(\xi)-f(x_2)\}| \\ &< \epsilon.\end{aligned}$$

In either case $|f(x')-f(x'')|<\epsilon$ whenever $|x'-x''|<\delta_0$, where δ_0 is *independent* of x', x''; that is to say, $f(x)$ is uniformly continuous.

13. Proofs of the following theorems may be left to the reader.

I. If AB be a simple curve with limited variations and if $f(z)$ be continuous on the curve AB, then

$$\int_A^B f(z)\,dz = -\int_B^A f(z)\,dz.$$

That is to say, changing the orientation of the path of integration changes the sign of the integral of a given function.

II. If C be a point on the simple curve AB, and if $f(z)$ be continuous on the curve, then

$$\int_A^B f(z)\,dz = \int_A^C f(z)\,dz + \int_C^B f(z)\,dz.$$

III. If z_0 and Z be the complex coordinates of A and B respectively, and if AB be a simple curve joining A, B, then

$$\int_A^B dz = Z - z_0.$$

IV. With the notation of Theorems I and II of § 11, by taking $z_p = z_p^{(n)}$, and Z_p in turn equal to $z_p^{(n)}$ and $z_{p+1}^{(n)}$, it follows that

$$\begin{aligned}\int_A^B z\,dz &= \tfrac{1}{2}\lim_{n\to\infty}\sum_{r=0}^{n}\left[(z_{r+1}^{(n)}-z_r^{(n)})\,z_r^{(n)}\right] \\ &\qquad + \tfrac{1}{2}\lim_{n\to\infty}\sum_{r=0}^{n}\left[(z_{r+1}^{(n)}-z_r^{(n)})\,z_{r+1}^{(n)}\right] \\ &= \tfrac{1}{2}\lim_{n\to\infty}\sum_{r=0}^{n}\left[(z_{r+1}^{(n)})^2-(z_r^{(n)})^2\right] \\ &= \tfrac{1}{2}(Z^2 - z_0^2).\end{aligned}$$

CHAPTER III

CAUCHY'S THEOREM

14. Let C_1, C_2 be two unclosed simple curves with the same end-points, but no other common points, each curve having limited variations. If z_0, Z be the end-points and if $f(z)$ be a function of z which is continuous on each curve and is one-valued at z_0 and Z, then

$$\int_{C_1} f(z)\,dz, \quad \int_{C_2} f(z)\,dz$$

both exist.

If $f(z)=z$, it follows from Theorem IV of § 13 that these two integrals have the same value. Further, if C_1, C_2 be oriented so that z_0 is the first point of C_1 and Z the first point of C_2, and if C_1, C_2 have no points in common save their end-points, C_1 and C_2 taken together form a simple closed curve, C, with limited variations, and

$$\int_C z\,dz = 0.$$

This result suggests that the circumstances in which

$$\int_C f(z)\,dz = 0,$$

(where C denotes a simple closed curve with limited variations[1] and $f(z)$ denotes a function of z which is continuous on C) should be investigated.

[1] A regular closed curve, satisfying this condition, regarded as a path of integration, is usually described as a *closed contour*.

The investigation appears all the more necessary from the fact[2] that if C be the unit circle $|z|=1$, described counterclockwise, and $f(z)=z^{-1}$, (so that $z=\cos t+i\sin t$, $-\pi \leqslant t \leqslant \pi$), it can be shewn that

$$\int_C z^{-1}dz = 2\pi i.$$

Conditions for the truth of the equation

$$\int_C f(z)\,dz = 0$$

were first investigated by Cauchy[3].

It is *not sufficient* that $f(z)$ should be continuous and one-valued on the regular closed curve C, as is obvious from the example cited, in which $f(z)=z^{-1}$, and, further, it is *not sufficient* that $f(z)$ should be continuous at all points of C and its interior.

A sufficient condition for the truth of the equation is that, given a function $f(z)$ which exists and is continuous and one-valued on the curve C, it should be possible to define a function[4], $f(z)$, which exists and is continuous and is one-valued at all points of the closed region formed by C and its interior, and which possesses the further property that the unique limit

$$\lim_{\substack{x'\to x\\ y'\to y}} \frac{f(z')-f(z)}{z'-z}$$

should exist at every point z of this closed region, it being supposed that z' is a point of the closed region The existence of this limit implies the continuity of $f(z)$ in the region.

It is, further, convenient, in setting out the proof, to lay a restriction on the contour C, namely that if a line be drawn parallel to Ox or to Oy, the portions of the line which are not points of C form a finite number of segments. This restriction will be removed in § 17.

15. Definition. *Analytic functions.* The one-valued continuous function $f(z)$ is said to be analytic at a point z of a continuum, if a number, l, can be found satisfying the condition that, given an

[2] Hardy, *A Course of Pure Mathematics*, § 204.

[3] *Mémoire sur les intégrales définies prises entre des limites imaginaires* (1825); this memoir is reprinted in t. VII. and t. VIII of the *Bulletin des Sciences Mathématiques.*

[4] Up to the present point a function, $f(z)$, of the complex variable z, has meant merely a function of the two real variables x and y.

arbitrary positive number ϵ, it is possible to find a positive number δ (depending on ϵ and z) such that

$$|f(z') - f(z) - l(z' - z)| \leqslant \epsilon |(z' - z)|,$$

for *all* values of z' such that $|z' - z| \leqslant \delta$.

The number l is called the differential coefficient, or derivate, of $f(z)$; if we regard z as variable, l is obviously a function of z; we denote the dependence of l upon z by writing $l = f'(z)$.

So far as Cauchy's theorem, that $\int_C f(z)\, dz = 0$, is concerned, it is not necessary that $f(z)$ should be analytic at points actually on C; it is sufficient that $f(z)$ should be analytic at all points of the interior of C and that for every point, z, of C,

$$|f(z') - f(z) - f'(z) . (z' - z)| \leqslant \epsilon |z' - z|,$$

whenever $|z' - z| < \delta$ (where δ depends on ϵ and z), provided that z' is a point of the closed region formed by C and its interior.

In such circumstances, we shall say that $f(z)$ is *semi-analytic on C*.

It is not difficult to see that analytic functions form a more restricted class than continuous functions. The existence of a unique differential coefficient implies the continuity of the function; whereas the converse is not true; for e.g. $|z|$ is continuous but not analytic.

16. It is now possible to prove CAUCHY'S THEOREM, namely that:

If $f(z)$ be analytic at all points in the interior of a regular closed curve with limited variations, C, and if the function be continuous throughout the closed region formed by C and its interior, then

$$\int_C f(z)\, dz = 0.$$

The theorem will first be proved on the hypothesis that $f(z)$ is subject to the further restriction that it is to be semi-analytic on C.

In accordance with § 8, let the orientation of C be determined in the conventional manner, so that if the (coincident) end-points of the path of integration be called z_0 and Z, with parameters t_0 and T, then, as t increases from t_0 to T, z describes C in the counterclockwise direction.

The continuum formed by the interior of C will be called R^-; and the closed region formed by R^- and C will be called R.

Let L be the sum of the variations of x and y as z describes the curve C; take any point of R^-, and with it as centre describe a square of side $2L$, the sides of the square being parallel to the axes; then no point of R lies outside this square; for if (x_0, y_0) be the centre of the square and x_1, x_2 the extreme values of x on R, then

$$x_1 \leqslant x_0 \leqslant x_2, \quad 0 \leqslant x_2 - x_1 \leqslant L,$$

so that $x_0 + L \geqslant x_1 + L \geqslant x_2$; i.e. the right-hand side of the square is on the right of R; applying similar reasoning to the other three sides of the square, it is apparent that no point of R is outside the square.

Let ϵ be an arbitrary positive number, then, since $f(z)$ is analytic inside C and semi-analytic on C, corresponding to any point, z, of R we can find a positive number δ such that

$$|f(z') - f(z) - (z' - z) f'(z)| \leqslant \epsilon |z' - z|,$$

whenever $|z' - z| \leqslant \delta$ and z' is a point of R.

Hence, by Goursat's lemma (§ 12), we can divide R into a finite number of sets of points such that a point, z_1, of each set can be found such that

$$|f(z') - f(z_1) - (z' - z_1) f'(z_1)| \leqslant \epsilon |z' - z_1|,$$

where z' is any member of the set to which z_1 belongs.

Suppose that R is divided into such sets, as in the proof of Goursat's lemma, by the process of dividing up the square of side $2L$ into four equal squares, and repeating the process of dividing up any of these squares into four equal squares, if such a process is necessary.

The effect of bisecting the square of side $2L$ is to divide R^- into a *finite* number of continua, by Theorem VI of § 6 combined with the hypothesis at the end of § 14; the boundaries of these continua are C and the straight line which bisects the square; the process of dividing up the square again is to divide these continua into other continua; and finally when R has been divided into suitable sets, R^- has been divided into a finite number of continua whose boundaries are portions of C and portions of the sides of the squares.

The squares into which the square of side $2L$ has been divided fall into the following three classes:

(i) Squares such that every point inside them is a point of R.

(ii) Squares such that some points inside them are points of R, but other points inside them are not points of R.

(iii) Squares such that no point inside them is a point of R.

The points inside C which are inside any particular square of class (i) form a continuum, namely the interior of the square; the points inside C which are inside any particular square of class (ii) form one or more continua.

Let the squares of class (i) be numbered from 1 to N and let the oriented boundary of the kth of these squares be called C_k.

Let the squares of class (ii) be numbered from 1 to N'. Let the set of oriented boundaries of the continua formed by points of R^- inside the kth of these squares be called C_k'.

Consider $$\sum_{k=1}^{N} \int_{(C_k)} f(z)\,dz + \sum_{k=1}^{N'} \int_{(C_k')} f(z)\,dz;$$
we shall shew that this sum is equal to $\int_{(C)} f(z)\,dz$.

The interiors of the squares of class (i), and the interiors of the regions whose complete boundaries are C_k', are all mutually external The boundaries formed by all those parts of the sides of the squares which belong to R^- occur twice in the paths of integration, and the whole of the curve C occurs once in the path of integration. By Theorem II of § 8, each path of integration which occurs twice in the sum occurs with opposite orientations; so that the integrals along these paths cancel, by Theorem I of § 13.

Again, the interiors of all the regions whose boundaries are C_k and C_k' are interior to C; so that the orientation of each part of C which occurs in the paths of integration is the same as the orientation of C; and therefore the paths of integration which occur once in the summation add up to produce the path of integration C (taken counter-clockwise). Consequently
$$\sum_{k=1}^{N} \int_{(C_k)} f(z)\,dz + \sum_{k=1}^{N'} \int_{(C_k')} f(z)\,dz = \int_{(C)} f(z)\,dz.$$

Now consider $\int_{(C_k)} f(z)\,dz$; the closed region formed by the square C_k and its interior has been chosen in such a way that a point z_1 of the region can be found such that
$$|f(z) - f(z_1) - (z - z_1) f'(z_1)| < \epsilon |(z - z_1)|,$$
when z is any point of the region.

Let $$f(z) - f(z_1) - (z - z_1) f'(z_1) = v(z - z_1),$$
when $$z \neq z_1.$$

When $z = z_1$ let $v = 0$; then v is a function of z and z_1 such that $|v| < \epsilon$. It follows that
$$\int_{(C_k)} f(z)\,dz = \int_{(C_k)} \{f(z_1) - z_1 f'(z_1)\}\,dz + \int_{(C_k)} z f'(z_1)\,dz + \int_{(C_k)} (z - z_1)\,v\,dz.$$

But by Theorem III of § 13, $\int dz = Z - z_0$, where z_0, Z are the end-points of the path of integration; since C_k is a closed curve, $Z = z_0$, so that $\int_{(C_k)} dz = 0$; so also, by Theorem IV of § 13, $\int_{(C_k)} z\,dz = 0$.

Therefore $$\int_{(C_k)} f(z)\,dz = \int_{(C_k)} (z - z_1)\,v\,dz.$$

Therefore[5], since the modulus of a sum is less than or equal to the sum of the moduli,

$$\begin{aligned} \left|\int_{(C_k)} f(z)\,dz\right| &= \left|\int_{(C_k)} (z - z_1)\,v\,dz\right| \\ &\leqslant \int_{(C_k)} |(z - z_1)\,v\,dz| \\ &\leqslant \int_{C_k} l_k\sqrt{2}\,\epsilon\,|dz| \\ &\leqslant l_k\,\epsilon\sqrt{2}\,.\,4l_k \\ &\leqslant 4\epsilon A_k\sqrt{2}, \end{aligned}$$

where l_k is the side[6] of C_k and A_k is the area of C_k, so that $A_k = l_k^2$; it is obvious by the lemma of § 9 that $\int_{(C_k)} |dz|$ does not exceed the perimeter of C_k.

We next consider $\int_{(C_k')} f(z)\,dz$, if the region of which C_k' is the total boundary consists of more than one continuum (i.e. if C_k' consists of more than one regular closed curve), we regard C_k' as being made up of a finite number of regular closed curves; and since the interior of each of these lies wholly inside C, any portion of any of them which coincides with a portion of C has the same orientation as C; and the value of $\int dz$, $\int z\,dz$ round each of the regular closed curves which make up C_k' is zero.

Hence, as in the case of C_k, we get

$$\begin{aligned} \left|\int_{(C_k')} f(z)\,dz\right| &= \left|\int_{(C_k')} v\,(z - z_1)\,dz\right| \\ &\leqslant \int_{(C_k')} (l_k'\sqrt{2})\,.\,\epsilon\,|dz|, \end{aligned}$$

[5] The expression $\int |f(z)\,dz|$ means $\lim_{n\to\infty} \sum_{r=0}^{n} |(z_{r+1}^{(n)} - z_r^{(n)}) f(z_r^{(n)})|$, with the notation of Chapter II; arguments similar to those of Chapter II shew that the limit exists.

[6] The squares C_k are not necessarily of the same size.

where l_k' is the length of the side of that square of class (ii) in which C_k' lies.

Let the sum of the variations of x and y, as z describes the portions of C which lie on C_k', be L_k'; so that

$$\sum_{k=1}^{N'} L_k' \leqslant L.$$

$\left(\sum_{k=1}^{N'} L_k'\right.$ will be less than L if part of C coincides with a portion of a side or sides of squares of class (i).$\left.\right)$

Now $$\int_{(C_k')} |dz| \leqslant L_k' + 4l_k';$$

for, by the lemma of § 9, $\int_{(C_k')} |dz|$ is less than or equal to the sum of the variations of x and y as z describes the various portions of C_k'.

Therefore $$\left|\int_{(C_k')} f(z)\, dz\right| \leqslant (L_k' + 4l_k')\,\epsilon\, l_k' \sqrt{2}$$
$$\leqslant 4\epsilon A_k' \sqrt{2} + 2L\epsilon L_k' \sqrt{2},$$

since $l_k' \leqslant 2L$; A_k' is the area of the square C_k'.

Combining the results obtained, it is evident that

$$\left|\int_C f(z)\, dz\right| = \left|\left\{\sum_{k=1}^{N} \int_{(C_k)} f(z)\, dz + \sum_{k=1}^{N'} \int_{(C_k')} f(z)\, dz\right\}\right|$$
$$\leqslant \sum_{k=1}^{N} \left|\int_{(C_k)} f(z)\, dz\right| + \sum_{k=1}^{N'} \left|\int_{(C_k')} f(z)\, dz\right|$$
$$\leqslant \sum_{k=1}^{N} 4A_k \epsilon \sqrt{2} + \sum_{k=1}^{N'} (4A_k' \epsilon \sqrt{2} + 2L\epsilon L_k' \sqrt{2}).$$

But it is evident that $\sum_{k=1}^{N} A_k + \sum_{k=1}^{N'} A_k'$ is not greater than the area of the square of side $2L$ which encloses C; and since $\sum_{k=1}^{N'} L_k' \leqslant L$, we see that

$$\left|\int_{(C)} f(z)\, dz\right| \leqslant 4 \times (2L)^2 \times \epsilon \sqrt{2} + 2\epsilon L^2 \sqrt{2}$$
$$\leqslant 18\epsilon L^2 \sqrt{2}.$$

Since L is independent of ϵ, the modulus of $\int_{(C)} f(z)\, dz$ is less than a number which we can take to be *arbitrarily small*. Hence $\int_{(C)} f(z)\, dz$ is zero, if $f(z)$ be analytic inside C and semi-analytic on C.

17. The results of the following two theorems make it possible to remove the restriction laid on C in § 14, namely that if a line be drawn parallel to Ox or to Oy, those portions of the line which are not points of C form a finite number of segments; also it will follow that the assumption made at the beginning of § 16, that $f(z)$ is semi-analytic on C, is unnecessary.

THEOREM I. *Given*[7] *a regular closed curve C and a positive number δ, a closed polygon D can be drawn such that every point of D is inside C and such that, given any point P on C, a point Q on D can be found such that $PQ < \delta$.*

THEOREM II. *If $f(z)$ be continuous throughout C and its interior, then $\int_C f(z)\,dz - \int_D f(z)\,dz$ can be made arbitrarily small by taking δ sufficiently small.*

It is obvious that the condition of § 14 is satisfied for polygons, so that if $f(z)$ be continuous throughout C and its interior and if it be analytic inside C, $\int_D f(z)\,dz=0$, and therefore $\int_C f(z)\,dz=0$.

THEOREM I. Let the elementary curves which form C be, in order,

$$y=g_1(x),\quad x=h_1(y),\quad y=g_2(x),\quad x=h_2(y),\ \ldots\ y=g_s(x),\quad x=h_s(y),$$

and let the interior of C be called S^-.

Let
$$\delta < \lim\sup \tfrac{1}{2}PQ,$$
where P, Q are any two points on C.

Each of the elementary curves which form C can be divided into a finite number of segments such that the sum of the fluctuations of x and y on each segment does not exceed $\frac{1}{4}\delta$, so that $\lim\sup PQ \leqslant \frac{1}{4}\delta$, where P, Q are any two points on one segment. Let each elementary curve be divided into at least three such segments and let the segments taken in order on C be called $\sigma_1, \sigma_2, \ldots \sigma_{n+1}$, their end-points being called $P_0, P_1, \ldots P_{n+1} (=P_0)$.

Choose $\delta' \leqslant \delta$ so that $\lim\inf PQ > \delta'$, where P, Q are any two points of C which do not lie on the same or on adjoining segments[8].

Cover the plane with a network of squares whose sides are parallel to the axes and of length $\frac{1}{4}\delta'$; if the end-point of any segment σ_r lies on the side of a square, shift the squares until this is no longer the case.

Take all the squares which have any point of σ_r inside or on them; these squares form a single closed region S_r; for if σ_r be on $y=g(x)$, the squares forming S_r can be grouped in columns, each column abutting on the column on its left and also on the column on its right. Let the boundary and interior of S_r be called C_r and S_r^- respectively.

Then S_r possesses the following properties:

(i) S_r contains points inside C and points outside C.

[7] This result will be obtained by the methods of de la Vallée Poussin, *Cours d'Analyse Infinitésimale* (1914), §§ 343–344.

[8] See note 15, p. 10.

(ii) S_r^- has at least one point P_r (and therefore the interior of one square) in common with S^-_{r+1}.

(iii) S_r, S_{r+2} have no point in common; for if they had a common point P, points Q_r, Q_{r+2} could be found on σ_r, σ_{r+2} respectively, such that $PQ_r \leqslant \frac{1}{4}\delta' \surd 2$, $PQ_{r+2} \leqslant \frac{1}{4}\delta' \surd 2$, and then $Q_r Q_{r+2} \leqslant \frac{1}{2}\delta' \surd 2 < \delta'$, which is impossible.

(iv) Since S_{r-1}, S_{r+1} have no common point, S_r consists of at least three squares.

(v) If $y=g(x)$ has points on m squares which lie on a column, the sum of the fluctuations of x and y as the curve completely crosses the column is at least $(m-1)\delta'$, (or δ' if $m=1$); in the case of a column which the curve does not completely cross, the sum of the fluctuations is at least $(m-2)\delta'$, (or 0 if $m=1$). The reader will deduce without much difficulty that the ratio of the perimeter of S_r to the sum of the fluctuations of x and y on σ_r cannot exceed 12; in the figure, the ratio is just less than 12 for the segment σ_{r+1}.

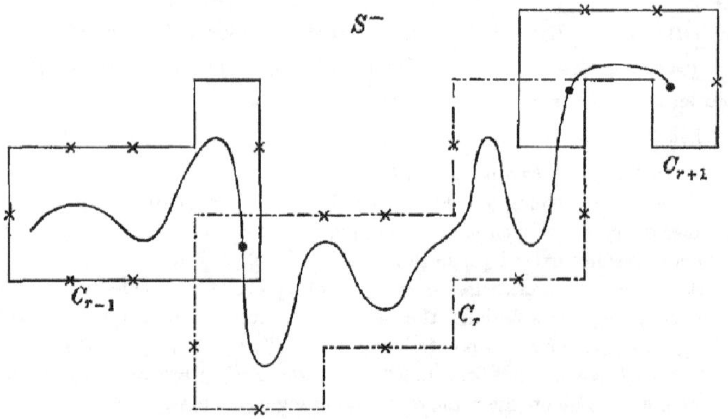

If σ_{r-1}, σ_r, σ_{r+1} be all on the same elementary curve, it is easy to see that a point describing C_r counter-clockwise (starting at a point inside C and outside C_{r-1}, C_{r+1}) will enter S^-_{r-1}, emerge from S^-_{r-1} outside C, enter S^-_{r+1} outside C and then emerge from S^-_{r+1}.

If, however, σ_{r-1}, σ_r be on adjacent elementary curves, a point describing C_r may enter and emerge from S^-_{r-1} more than once; but it is possible to take a number of squares forming a closed region S_r', whose boundary is E_r, consisting of the squares of S_r and S_{r-1} together with the squares which lie in the regions (if any) which are completely surrounded by the squares of S_r and S_{r-1} Then, as a point describes E_r counter-clockwise, it enters and emerges from S^-_{r-2} and S^-_{r+1} only once. If we thus modify those regions S_r which correspond to end segments of the elementary curves, we get a set of $m+1$ $(< n)$ closed regions T_p, with boundaries D_p and interiors T_p^-, such that D_p meets D_{p+1} but non-consecutive regions are wholly external to one another.

Now consider the arc of each polygon D_p which lies outside $T^-{}_{p-1}$ and $T^-{}_{p+1}$ but inside C; these overlapping arcs form a closed polygon D which is wholly inside C, with arcs of D_1, D_2, .. , occurring on it in order. Also, if P be any point of σ_r, there is a point Q of σ_r or σ_{r+1} which is inside a square which abuts on D, and therefore the distance of P from some point of D does not exceed $PQ+\frac{1}{4}\delta' < \frac{1}{2}\delta+\frac{1}{4}\delta' < \delta$.

Theorem I is therefore completely proved.

THEOREM II. Let ϵ be an arbitrary positive number.

(i) Choose δ so small that

$$|f(z')-f(z)| \leqslant \tfrac{1}{32}\epsilon L_1^{-1},$$

whenever $|z'-z| \leqslant 3\delta$ and z, z' are any two points on or inside C, while $L_1=12L$, where L is the sum of the fluctuations of x and y on C.

(ii) Choose such a parameter t for the curve C that

$$|f(z')-f(z)| \leqslant \tfrac{1}{32}\epsilon L^{-1},$$

whenever $|t'-t| \leqslant \delta$; this is obviously possible, for, if the inequality were only true when $|t'-t| \leqslant \lambda\delta$, where λ is a positive number less than unity and independent of t, we should take a new parameter $\tau=\lambda^{-1}t$.

It is evident from (i) that

$$|f(z')-f(z)| \leqslant \tfrac{1}{32}\epsilon L^{-1},$$

whenever $|z'-z| \leqslant \delta$ and z, z' are any two points on C.

Draw the polygon D for the value of δ under consideration, as in Theorem I. Take any one of the curves D_p; if it wholly contains more than one of the regions S_n, let them be S_{r-1}, S_r. Then there is a point z_p of σ_{r-1} or σ_r in one of the squares of D_p which abuts on D; let ζ_p be a point on the side of this square which is part of D.

Then z_{p+1} is on σ_{r+1}, and hence $|z_{p+1}-z_p|$ does not exceed the sum of the fluctuations of x and y on σ_{r-1}, σ_r, σ_{r+1}; i.e. $|z_{p+1}-z_p| < \frac{3}{4}\delta < \delta$.

Also the arc of D joining ζ_p to ζ_{p+1} does not exceed 12 times the sum of the fluctuations of x and y on the arcs σ_{r-1}, σ_r, σ_{r+1} and so does not exceed 3δ; and the sum of the fluctuations of ξ, η as ζ describes D does not exceed the sum of the perimeters of the curves C_r, i.e. it does not exceed $L_1=12L$.

Take as the parameter τ, of a point ζ on D, the arc of D measured from a fixed point to ζ.

We can now consider the value of $\int_C f(z)\,dz$.

By conditions (i) and (ii) coupled with Theorem II of § 11, we see that, since $|\zeta_{p+1}-\zeta_p| \leqslant |\tau_{p+1}-\tau_p| \leqslant 3\delta$,

$$\left|\left\{\int_D f(\zeta)\,d\zeta - \sum_{p=0}^{m}(\zeta_{p+1}-\zeta_p)f(\zeta_p)\right\}\right| \leqslant \tfrac{1}{8}\epsilon,$$

and

$$\left|\left\{\int_C f(z)\,dz - \sum_{p=0}^{m}(z_{p+1}-z_p)f(z_p)\right\}\right| \leqslant \tfrac{1}{8}\epsilon.$$

But $\left|\int_C f(z)\,dz - \int_D f(\zeta)\,d\zeta\right|$

$$\leqslant \left|\left\{\int_C f(z)\,dz - \sum_{p=0}^{m}(z_{p+1}-z_p)f(z_p)\right\}\right|$$

$$+\left|\left\{\int_D f(\zeta)\,d\zeta - \sum_{p=0}^{m}(\zeta_{p+1}-\zeta_p)f(\zeta_p)\right\}\right|$$

$$+\left|\left\{\sum_{p=0}^{m}(z_{p+1}-z_p)f(z_p) - \sum_{p=0}^{m}(\zeta_{p+1}-\zeta_p)f(\zeta_p)\right\}\right|$$

$$\leqslant \tfrac{1}{4}\epsilon + \left|\sum_{p=0}^{m}\{(z_{p+1}-z_p)f(z_p)-(\zeta_{p+1}-\zeta_p)f(\zeta_p)\}\right|.$$

Write $$f(\zeta_p)=f(z_p)+v_p,\quad \zeta_p=z_p+\eta_p,$$

so that $$|v_p| < \tfrac{1}{32}\epsilon L_1^{-1},\quad |\eta_p| < \delta'.$$

Then

$$\left|\sum_{p=0}^{m}(z_{p+1}-z_p)f(z_p)-(\zeta_{p+1}-\zeta_p)f(\zeta_p)\right|$$

$$=\left|\sum_{p=0}^{m}[(\eta_p-\eta_{p+1})f(z_p)-(\zeta_{p+1}-\zeta_p)\{f(\zeta_p)-f(z_p)\}]\right|$$

$$=\left|\sum_{p=0}^{m}[\eta_{p+1}\{f(z_{p+1})-f(z_p)\}-(\zeta_{p+1}-\zeta_p)v_p]\right|$$

$$\leqslant \sum_{p=0}^{m}|\eta_{p+1}\{f(z_{p+1})-f(z_p)\}| + \sum_{p=0}^{m}|(\zeta_{p+1}-\zeta_p)v_p|.$$

Now $$\sum_{p=0}^{m}|\eta_{p+1}\{f(z_{p+1})-f(z_p)\}| < \tfrac{1}{32}(m+1)\delta'\epsilon L^{-1},$$

by condition (ii), while

$$\sum_{p=0}^{m}|(\zeta_{p+1}-\zeta_p)v_p| < \tfrac{1}{32}\epsilon L_1^{-1}\sum_{p=0}^{m}|\zeta_{p+1}-\zeta_p|$$

$$< \tfrac{1}{32}\epsilon.$$

Therefore, collecting the results and noticing that $(m+1)\delta < L$, we see that

$$\left|\int_C f(z)\,dz - \int_D f(z)\,dz\right| \leqslant \tfrac{1}{4}\epsilon + \tfrac{3}{8}\epsilon + \tfrac{1}{32}\epsilon$$

$$< \epsilon.$$

If now, in addition to the hypothesis of the enunciation of Theorem II, that $f(z)$ is continuous throughout C and its interior, we assume that $f(z)$ is analytic in the interior of C, then $f(z)$ is analytic throughout D and its interior, and so $\int_D f(z)\,dz=0$, by § 16; and then, by the result that $\left|\int_C f(z)\,dz\right| < \epsilon$, we infer that $\int_C f(z)\,dz=0$. The result stated at the beginning of § 16 has now been completely proved.

CHAPTER IV

MISCELLANEOUS THEOREMS

§ 18. Change of variable in an integral.—§ 19. Differentiation of an integral with regard to one of the limits.—§ 20. Uniform differentiability implies a continuous differential coefficient, and the converse.

18 *Change of variable in an integral.* Let ζ be the complex coordinate of any point on a simple curve AB, with limited variations. Let $z = g(\zeta)$ be a function of ζ which has a continuous differential coefficient, $g'(\zeta)$, at all points of the curve, so that, if ζ be any particular point of the curve, given a positive ϵ, we can find δ such that

$$|g(\zeta') - g(\zeta) - (\zeta' - \zeta)\, g'(\zeta)| \leqslant \epsilon\, |\zeta' - \zeta|,$$

when $|t' - t| \leqslant \delta$; it being supposed that t, t' are the parameters of ζ, ζ'.

If t_0, T be the parameters of A, B, suppose that z describes a simple curve CD as t increases from t_0 to T.

Then the equation

$$\int_{AB} f(g(\zeta))\, g'(\zeta)\, d\zeta = \int_{CD} f(z)\, dz$$

is true if $f(z)$ be a continuous function on the curve CD.

By Theorem II of § 11, given any positive number ϵ, it is possible to find a positive number δ' such that if any ν numbers $t_1, t_2, \ldots t_\nu$ are taken so that $0 \leqslant t_{p+1} - t_p \leqslant \delta'$, and if T_p be such that $t_p \leqslant T_p \leqslant t_{p+1}$, then

$$\left| \int_{CD} f(z)\, dz - \sum_{p=0}^{\nu} (z_{p+1} - z_p) f(Z_p) \right| < \epsilon.$$

Given the same number ϵ, we can find δ'' such that if any ν numbers $t_1, t_2, \ldots t_\nu$ are taken so that $0 \leqslant t_{p+1} - t_p \leqslant \delta''$, and if T_p be such that $t_p \leqslant T_p \leqslant t_{p+1}$, then

$$\left| \int_{AB} f(g(\zeta))\, g'(\zeta)\, d\zeta - \sum_{p=0}^{\nu} (\zeta_{p+1} - \zeta_p) f(Z_p)\, g'(W_p) \right| < \epsilon,$$

where W_p, Z_p are corresponding points on AB, CD; we take δ to be the smaller of δ', δ'' and choose the same values for $t_1, t_2, \ldots t_\nu$ in both summations, where $0 \leqslant t_{p+1} - t_p \leqslant \delta$, and we take T_p the same in both summations.

Now divide the range t_0 to T into any number of intervals each interval being less than δ; and subdivide each of these into a number of intervals which are 'suitable' for the inequality

$$|g(\zeta') - g(\zeta) - (\zeta' - \zeta)\, g'(\zeta)| \leqslant \epsilon(\zeta' - \zeta).$$

Then taking the end-points of these intervals to be $t_0, t_1, \ldots t_\nu, T$, and, taking T_p to be the point of the pth interval such that

$$|g(\zeta) - g(W_p) - (\zeta - W_p)\, g'(W_p)| \leqslant \epsilon\, |\zeta - W_p|$$

at all points ζ of the arc $\zeta_p \zeta_{p+1}$ of AB, we have

$$\left|\sum_{p=0}^{\nu} (z_{p+1} - z_p) f(Z_p) - \sum_{p=0}^{\nu} (\zeta_{p+1} - \zeta_p) f(Z_p)\, g'(W_p)\right|$$

$$= \left|\sum_{p=0}^{\nu} f(Z_p)\{g(\zeta_{p+1}) - g(\zeta_p) - (\zeta_{p+1} - \zeta_p)\, g'(W_p)\}\right|$$

$$= \left|\sum_{p=0}^{\nu} f(Z_p)\,[\{g(\zeta_{p+1}) - g(W_p) - (\zeta_{p+1} - W_p)\, g'(W_p)\} - \{g(\zeta_p) - g(W_p) - (\zeta_p - W_p)\, g'(W_p)\}]\right|$$

$$\leqslant \sum_{p=0}^{\nu} |f(Z_p)\, \epsilon\, \{|\zeta_{p+1} - W_p| + |W_p - \zeta_p|\}|$$

Let L be the sum of the fluctuations of ξ, η on AB and let ML^{-1} be the upper limit of $|f(z)|$ on CD; M exists since $f(z)$ is continuous.

Then, by the last inequality,

$$\left|\sum_{p=0}^{\nu} (z_{p+1} - z_p) f(Z_p) - \sum_{p=0}^{\nu} (\zeta_{p+1} - \zeta_p) f(Z_p)\, g'(W_p)\right| < \epsilon M.$$

Therefore

$$\left|\int_{AB} f(g(\zeta))\, g'(\zeta)\, d\zeta - \int_{CD} f(z)\, dz\right|$$

$$= \left|\int_{AB} f(g(\zeta))\, g'(\zeta)\, d\zeta - \sum_{p=0}^{\nu} (\zeta_{p+1} - \zeta_p) f(Z_p)\, g'(W_p) - \int_{CD} f(z)\, dz + \sum_{p=0}^{\nu} (z_{p+1} - z_p) f(Z_p) + \sum_{p=0}^{\nu} (\zeta_{p+1} - \zeta_p) f(Z_p)\, g'(W_p) - \sum_{p=0}^{\nu} (z_{p+1} - z_p) f(Z_p)\right|$$

$$< (2 + M)\, \epsilon;$$

since M is fixed, ϵ is arbitrarily small and the two integrals exist, we infer that

$$\int_{AB} f(g(\zeta))\, g'(\zeta)\, d\zeta = \int_{CD} f(z)\, dz.$$

Corollary. Taking $f(z)=1$, we see that

$$\int_{AB} g'(\zeta)\, d\zeta = \int_{CD} dz = z_D - z_C = g(\zeta_B) - g(\zeta_A);$$

this is the formula for the integral of a continuous differential coefficient.

19. *Differentiation of an integral with regard to one of the limits.*

Let AB be a regular unclosed curve such that if any point P on it be taken, and if Q be any other point of it, the ratio of the sum of the variations of the curve between P and Q to the length of the chord PQ has a finite upper limit[1], k.

Let $f(z)$ be continuous on the curve and let z_0, Z, $Z+h$ be any three points on it; then if z_0 be fixed, $\int_{z_0}^{Z} f(z)\, dz$ is a function of Z only, say $\phi(Z)$; and

$$\lim_{t\to 0} \frac{\phi(Z+h)-\phi(Z)}{h} = f(Z),$$

where t is the difference of the parameters of Z, $Z+h$.

We can find δ so that $|f(Z+h)-f(Z)| < \epsilon$ when $t<\delta$, where ϵ is arbitrary.

Now

$$\begin{aligned} h^{-1}\{\phi(Z+h)-\phi(Z)\} &= h^{-1}\int_{Z}^{Z+h} f(z)\, dz \\ &= h^{-1} \lim_{n\to\infty} \sum_{r=1}^{n} f(Z+h_r^{(n)}) \,.\, (h_{r+1}^{(n)} - h_r^{(n)}), \end{aligned}$$

where $h_0^{(n)}=0$, $h_{n+1}^{(n)}=h$; it being supposed that the points $h_r^{(n)}$ are chosen in the same way as the points $z_r^{(n)}$ in § 10 of Chapter II.

Therefore

$$\begin{aligned} &|h^{-1}\{\phi(Z+h)-\phi(Z)\}-f(Z)| \\ &= \left| h^{-1} \lim_{n\to\infty} \sum_{r=1}^{n} f(Z+h_r^{(n)}) \,.\, (h_{r+1}^{(n)}-h_r^{(n)}) - h^{-1} f(Z) \sum_{r=1}^{n} (h_{r+1}^{(n)}-h_r^{(n)}) \right| \\ &= \left| h^{-1} \lim_{n\to\infty} \sum_{r=1}^{n} \{f(Z+h_r^{(n)})-f(Z)\} \,.\, (h_{r+1}^{(n)}-h_r^{(n)}) \right| \end{aligned}$$

[1] This condition is satisfied by most curves which occur in practice.

$$\leqslant |h^{-1}| \lim_{n\to\infty} \sum_{r=1}^{n} |\{f(Z+h_r^{(n)}) - f(Z)\}| \,.\, |h_{r+1}^{(n)} - h_r^{(n)}|$$

$$< |h^{-1}| \,.\, \epsilon \sum_{r=1}^{n} |h_{r+1}^{(n)} - h_r^{(n)}|.$$

But, by the lemma of § 9,

$$|h^{-1}| \sum_{r=1}^{n} |h_{r+1}^{(n)} - h_r^{(n)}| < k,$$

so that
$$|h^{-1}\{\phi(Z+h) - \phi(Z)\} - f(Z)| < k\epsilon,$$
since ϵ is arbitrary and h is fixed, it follows from the definition of a limit that

$$\lim_{t\to 0} \frac{\phi(Z+h) - \phi(Z)}{h} = f(Z).$$

20. *Uniform differentiability implies a continuous differential coefficient, and the converse.*

Let $f(z)$ be uniformly differentiable throughout a region; so that when ϵ is taken arbitrarily, a positive number δ, independent of z, exists such that

$$|f(z') - f(z) - (z'-z) f'(z)| \leqslant \tfrac{1}{2}\epsilon |z'-z|,$$

whenever $|z'-z| \leqslant \delta$ and z, z' are two points of the region.

Since $|z-z'| \leqslant \delta$, we have

$$|f(z) - f(z') - (z-z') f'(z')| \leqslant \tfrac{1}{2}\epsilon |z-z'|.$$

Combining the two inequalities, it is obvious that

$$|(z'-z)\{f'(z') - f'(z)\}| \leqslant \tfrac{1}{2}\epsilon |z'-z| + \tfrac{1}{2}\epsilon |z-z'|,$$

and therefore
$$|f'(z') - f'(z)| \leqslant \epsilon,$$
whenever $|z'-z| \leqslant \delta$; that is to say, $f'(z)$ is continuous.

To prove the converse theorem, let $f'(z)$ be continuous, and therefore uniformly continuous, in a region; so that, when ϵ is taken arbitrarily, a positive number δ, independent of z, exists such that

$$|f'(z') - f'(z)| \leqslant \tfrac{1}{2}\epsilon,$$

whenever $|z'-z| \leqslant \delta$.

Consider only those points z whose distance from the boundary of the region exceeds δ; take $|z - Z| \leqslant \delta$.

Then since $f(z)$ is differentiable, to each point ζ of the straight line joining z to Z there corresponds a positive number δ_ζ such that

$$|f(\zeta') - f(\zeta) - (\zeta' - \zeta) f'(\zeta)| \leqslant \tfrac{1}{2}\epsilon |\zeta' - \zeta|,$$

whenever $|\zeta' - \zeta| \leqslant \delta_\zeta$ and ζ' is on the line zZ.

By Goursat's lemma, we may divide the line zZ into a *finite* number of intervals, say at the points $\zeta_0 (=z)$, ζ_1, $\zeta_2 \ldots \zeta_n$, $\zeta_{n+1} (=Z)$, such that there is a point z_r in the rth interval which is such that

$$|f(\zeta) - f(z_r) - (\zeta - z_r) f'(z_r)| \leqslant \tfrac{1}{2}\epsilon |\zeta - z_r|,$$

for all points ζ of the interval.

Therefore $\quad f(\zeta_r) - f(z_r) - (\zeta_r - z_r) f'(z_r) = v_r (\zeta_r - z_r),$

$$f(\zeta_{r-1}) - f(z_r) - (\zeta_{r-1} - z_r) f'(z_r) = v_r' (\zeta_{r-1} - z_r),$$

where $\quad |v_r| \leqslant \tfrac{1}{2}\epsilon, \ |v_r'| \leqslant \tfrac{1}{2}\epsilon.$

Also, since $\quad |z_r - z| \leqslant \delta,$

$$f'(z_r) = f'(z) + \eta_r,$$

where $\quad |\eta_r| \leqslant \tfrac{1}{2}\epsilon.$

Therefore $f(\zeta_r) - f(\zeta_{r-1}) - (\zeta_r - \zeta_{r-1}) f'(z_r)$

$$= \eta_r (\zeta_r - \zeta_{r-1}) + v_r (\zeta_r - z_r) - v_r' (\zeta_{r-1} - z_r).$$

Taking $r = 1, 2, \ldots n+1$ in turn, and summing we get

$$f(Z) - f(z) - (Z - z) f'(z)$$
$$= \sum_{r=1}^{n+1} \eta_r (\zeta_r - \zeta_{r-1}) + \sum_{r=1}^{n+1} \{v_r (\zeta_r - z_r) - v_r' (\zeta_{r-1} - z_r)\}.$$

But, since the points $\zeta_0 (=z)$, z_1, ζ_1, $z_2 \ldots \zeta_n$, z_n, $\zeta_{n+1} (=Z)$ are in order on a straight line,

$$\sum_{r=1}^{n+1} \{|(\zeta_r - z_r)| + |(\zeta_{r-1} - z_r)|\} = |Z - z|,$$

and so $\quad |f(Z) - f(z) - (Z - z) f'(z)| \leqslant \tfrac{1}{2}\epsilon |Z - z| + \tfrac{1}{2}\epsilon |Z - z|,$

whenever $|Z - z| \leqslant \delta$ and the distance of z from the boundary of the region does not exceed δ. Therefore, if $f'(z)$ is continuous throughout a region, $f(z)$ is uniformly differentiable throughout the interior of the region.

The reader will find no difficulty in proving the corresponding theorems when $f(z)$ is uniformly differentiable or when $f'(z)$ is continuous, and z is, in each case, restricted to be a continuous function of a real variable t.

CHAPTER V

THE CALCULUS OF RESIDUES

21. Let C be a closed contour and let $f(z)$ be a function of z which is continuous throughout C and its interior, and analytic inside C. Let a be the complex coordinate of any point P not on C. *Then the extension of Cauchy's theorem is that*

$$\left.\begin{aligned}\frac{1}{2\pi i}\int_C \frac{f(z)}{z-a}\,dz &= 0 \quad \text{if } P \text{ be outside } C \\ &= f(a) \text{ if } P \text{ be inside } C\end{aligned}\right\}.$$

The first part is almost obvious; for if P be outside C it is easily proved that $f(z)/(z-a)$ is analytic at points inside C and continuous on C. Therefore, by the result of Chapter III,

$$\frac{1}{2\pi i}\int_C \frac{f(z)}{z-a}\,dz = 0.$$

Now let P be a point inside C.

Through P draw a line parallel to Ox; there will be two[1] points Q_1, Q_2 on this line, one on the right of P, the other on the left, such that Q_1, Q_2 are on C, but no point of Q_1Q_2 except its end-points lies on C. [The existence of the points Q_1, Q_2 may be established by arguments similar to those in small print at the foot of page 11.]

[1] Points on the line which are sufficiently distant from P either to the right or left are outside C. Since a straight line is a simple curve, the straight lines joining P to these distant points meet C in one point at least.

Q_1, Q_2 divide C into two parts σ_1, σ_2 with the same orientations as C; let σ_1, σ_2 be chosen so that Q_1, Q_2 are the end-points of the oriented curve σ_1, and Q_2, Q_1 are the end-points of the oriented curve σ_2.

Let the shortest distance of points on C from P be δ_1; choose δ so that $\delta \leqslant \delta_1$, $\delta < 1$ and

$$|f(z) - f(a) - (z-a)f'(a)| \leqslant \epsilon |z-a|$$

when $|z-a| \leqslant \delta$, where ϵ is an arbitrary positive number; draw a circle with centre P and radius r ($\leqslant \frac{1}{2}\delta$).

Let $Q_1 Q_2$ meet this circle in P_1, P_2; let the upper half of the circle, with the orientation $(a-r, a+r)$ of its end-points, be called B_1, and let the lower half of the circle, with the orientation $(a+r, a-r)$ of its end-points, be called B_2.

Let the circle, properly oriented, be called C_3, so that the orientations of B_1 and B_2 are opposite to that of C_3.

Proofs of the following theorems are left to the reader:

(i) σ_1, $Q_2 P_2$, B_1, $P_1 Q_1$ form a closed contour, C_1, properly oriented.

(ii) σ_2, $Q_1 P_1$, B_2, $P_2 Q_2$ form a closed contour, C_2, properly oriented.

(iii) P is outside C_1 and C_2.

(iv) $f(z)/(z-a)$ is analytic inside C_1 and C_2; and it is continuous throughout the regions formed by C_1, C_2 and their interiors.

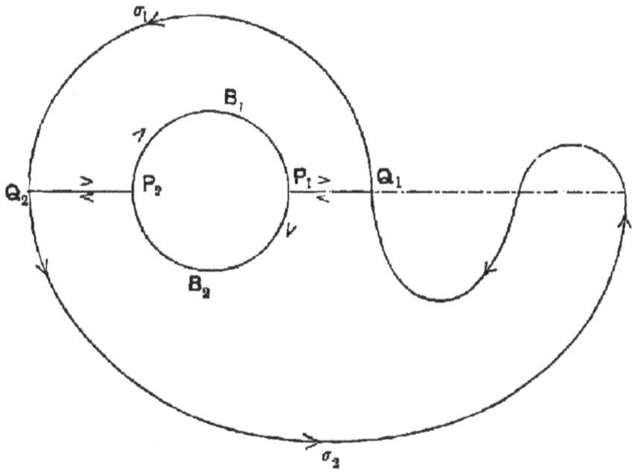

Now consider

$$\int_{C_1} \frac{f(z)}{z-a}\,dz + \int_{C_2} \frac{f(z)}{z-a}\,dz.$$

The path of integration consists of the oriented curves σ_1, σ_2, P_1Q_1, Q_1P_1, P_2Q_2, Q_2P_2, B_1, B_2.

The integrals along the oriented curves σ_1, σ_2 make up the integral along C. The integrals along the oriented curves P_1Q_1, Q_1P_1 cancel, and so do the integrals along the oriented curves P_2Q_2, Q_2P_2; while the integrals along the oriented curves B_1, B_2 make up minus the integral along the oriented curve C_3, since the orientations of B_1, B_2 are opposite to the orientation of C_3.

Hence

$$\int_{C_1}\frac{f(z)}{z-a}\,dz+\int_{C_2}\frac{f(z)}{z-a}\,dz=\int_{C}\frac{f(z)}{z-a}\,dz-\int_{C_3}\frac{f(z)}{z-a}\,dz.$$

Now $f(z)/(z-a)$ is analytic inside C_1 and C_2 and is continuous throughout the regions formed by C_1, C_2 and their interiors. Hence, by § 17, the integrals along C_1 and C_2 vanish.

Hence[2] $$\int_{C}\frac{f(z)}{z-a}\,dz=\int_{C_3}\frac{f(z)}{z-a}\,dz.$$

Let $$f(z)-f(a)-(z-a)f'(a)=v(z-a),$$
so that, when z is on C_3, $|v|\leqslant\epsilon$.

Then

$$\int_{C_3}\frac{f(z)}{z-a}\,dz=f(a)\int_{C_3}\frac{dz}{z-a}+f'(a)\int_{C_3}dz+\int_{C_3}v\,dz.$$

But, since C_3 is a closed curve, $\int_{C_3}dz=0$, by Theorem III of § 13.

To evaluate $\int_{C_3}\frac{dz}{z-a}$, put $z=a+r(\cos\theta+i\sin\theta)$; θ is a real number and is the angle which the line joining z to a makes with Ox. Consequently, since a is inside C_3, θ increases by 2π as z describes C_3.

Hence, by the result of § 18,

$$\int_{C_3}\frac{dz}{z-a}=\int_{\alpha}^{\alpha+2\pi}\frac{-\sin\theta+i\cos\theta}{\cos\theta+i\sin\theta}\,d\theta$$
$$=2\pi i.$$

Therefore $$\int_{C_3}\frac{f(z)}{z-a}\,dz-2\pi i f(a)=\int_{C_3}v\,dz,$$

[2] This result may be stated "The path of integration may be *deformed* from C into C_3 without affecting the value of the integral."

so that
$$\left|\int_C \frac{f(z)}{z-a}\,dz - 2\pi i f(a)\right| \leqslant \int_{C_3} |v\,dz|$$
$$\leqslant \epsilon \int_{C_3} |dz|.$$

Putting $z = a + r(\cos t + i \sin t)$, we get[3]
$$\int_{C_3} |dz| = \lim_{n\to\infty} \sum_{r=0}^{n} r\,|\{\cos t_{r+1}^{(n)} + i \sin t_{r+1}^{(n)} - \cos t_r^{(n)} - i \sin t_r^{(n)}\}|,$$
where
$$t_r^{(n)} < t_{r+1}^{(n)}, \quad t_0^{(n)} = t_0, \quad t_{n+1}^{(n)} = t_0 + 2\pi,$$
so that
$$\int_{C_3} |dz| = \lim_{n\to\infty} \sum_{r=0}^{n} 2r\,|\sin \tfrac{1}{2}(t_{r+1}^{(n)} - t_r^{(n)})|$$
$$\leqslant \lim_{n\to\infty} \sum_{r=0}^{n} 2r \,.\, \tfrac{1}{2}(t_{r+1}^{(n)} - t_r^{(n)})$$
$$\leqslant 2\pi r.$$

Hence
$$\left|\int_C \frac{f(z)}{z-a}\,dz - 2\pi i f(a)\right|$$
is less than $2\pi r\epsilon$, where $r \leqslant 1$ and ϵ is arbitrarily small. Consequently it must be zero; that is to say
$$\int_C \frac{f(z)}{z-a}\,dz = 2\pi i f(a).$$

22. Let C be a closed contour, and let $f(z)$ be a function of z which is analytic at all points inside C and continuous throughout C and its interior; let a be the complex coordinate of any point inside C.

Then $f(z)$ possesses unique differential coefficients of all orders at a; and
$$\frac{d^n f(a)}{da^n} = \frac{n!}{2\pi i}\int_C \frac{f(z)}{(z-a)^{n+1}}\,dz.$$

All points sufficiently near a are inside C; let δ be a positive number such that all points satisfying the inequality $|z-a| \leqslant 2\delta$ are inside C; and let h be *any* complex number such that $|h| \leqslant \delta$.

Then, by § 21,
$$f(a) = \frac{1}{2\pi i}\int_C \frac{f(z)}{z-a}\,dz,$$
$$f(a+h) = \frac{1}{2\pi i}\int_C \frac{f(z)}{z-a-h}\,dz.$$

[3] The notation of Chapter II is being employed.

Therefore

$$\frac{f(a+h)-f(a)}{h}-\frac{1}{2\pi i}\int_C \frac{f(z)}{(z-a)^2}\,dz=\frac{h}{2\pi i}\int_C \frac{f(z)}{(z-a)^2(z-a-h)}\,dz.$$

Now when z is on C,

$$|z-a|\geqslant 2\delta,\ |z-a-h|\geqslant \delta,\ \text{and}^4\ |f(z)|<K,$$

where K is a constant (independent of h and δ).

Hence, if L be the sum of the variations of x and y on C

$$\left|\int_C \frac{f(z)}{(z-a)^2(z-a-h)}\,dz\right| \leqslant \int_C \left|\frac{f(z)\,dz}{(z-a)^2(z-a-h)}\right| \leqslant \frac{KL}{4\delta^3}.$$

Therefore

$$\frac{f(a+h)-f(a)}{h}-\frac{1}{2\pi i}\int_C \frac{f(z)}{(z-a)^2}\,dz=v,$$

where $$|v|\leqslant |h|\,KL/(8\pi\delta^3).$$

Hence, as $h\to 0$, v tends to the limit zero.

Therefore $$\lim_{h\to 0}\frac{f(a+h)-f(a)}{h}$$

has the value $\frac{1}{2\pi i}\int_C \frac{f(z)}{(z-a)^2}\,dz$; that is to say,

$$\frac{d}{da}f(a)=\frac{1}{2\pi i}\int_C \frac{f(z)}{(z-a)^2}\,dz.$$

The higher differential coefficients may be evaluated in the same manner; the process which has just been carried out is the justification of 'differentiating with regard to a under the sign of integration' the equation

$$f(a)=\frac{1}{2\pi i}\int_C \frac{f(z)}{z-a}\,dz.$$

If we assume that $\frac{d^n f(a)}{da^n}$ exists and

$$\frac{d^n f(a)}{da^n}=\frac{n!}{2\pi i}\int_C \frac{f(z)}{(z-a)^{n+1}}\,dz \quad \ldots \quad \ldots\ldots\ldots(6),$$

[4] On C, the real and imaginary parts of $f(z)$ are continuous functions of a real variable, t, and a continuous function is bounded. See Hardy, *A Course of Pure Mathematics*, § 89, Theorem I.

a similar process will justify differentiating this equation with regard to a under the sign of integration, so that

$$\frac{d^{n+1}f(a)}{da^{n+1}} = \frac{(n+1)!}{2\pi i}\int_C \frac{f(z)}{(z-a)^{n+2}}\,dz \quad \text{............(6 a).}$$

But (6) is true when $n = 1$; hence by (6 a), (6) is true when $n = 2$; and hence, by induction, (6) is true for all positive integral values of n.

23. DEFINITIONS. *Pole. Residue.* Let $f(z)$ be continuous throughout a closed contour C and its interior, except at certain points $a_1, a_2, \ldots a_m$, *inside* C, and analytic at all points inside C except at $a_1, a_2, \ldots a_m$.

Let a function $\psi(z)$ exist which satisfies the following conditions:

(i) $\psi(z)$ is continuous throughout C and its interior, analytic at all points inside C.

(ii) At points on and inside C, with the exception of $a_1, a_2, \ldots a_m$,

$$f(z) = \psi(z) + \sum_{r=1}^{m} \phi_r(z) \quad \text{.....................(7),}$$

where $$\phi_r(z) = \frac{b_{1,r}}{z-a_r} + \frac{b_{2,r}}{(z-a_r)^2} + \ldots + \frac{b_{n_r,r}}{(z-a_r)^{n_r}}.$$

Then $f(z)$ is said to have a *pole* of order n_r at the point a_r; the coefficient of $(z-a_r)^{-1}$, viz. $b_{1,r}$, is called the *residue* of $f(z)$ at a_r.

It is evident by the result of Chapter III that

$$\int_C \psi(z)\,dz = 0;$$

so that, by (7), $$\int_C f(z)\,dz = \sum_{r=1}^{m} \int_C \phi_r(z)\,dz.$$

24. This last equation enables us to evaluate $\int_C f(z)\,dz$; for consider $\int_C \phi_r(z)\,dz$. The only point inside C at which $\phi_r(z)$ is not analytic is the point $z = a_r$. With centre a_r draw a circle C' of positive radius ρ, lying wholly inside C; then by reasoning precisely similar to that of § 21, we can deform the path of integration C into C' without affecting the value of the integral, so that

$$\int_C \phi_r(z)\,dz = \int_{C'} \phi_r(z)\,dz.$$

To evaluate this new integral, write

$$z = a_r + \rho(\cos\theta + i\sin\theta),$$

so that θ increases by 2π as z describes C'; as in § 21, if α be the initial value of θ,

$$\int_{C'} \phi_r(z)\,dz = \int_\alpha^{\alpha+2\pi} \phi_r(z)\,\rho(-\sin\theta + i\cos\theta)\,d\theta$$

$$= \sum_{n=1}^{n_r} b_{n,r}\,\rho^{1-n}\,i\int_\alpha^{\alpha+2\pi}\{\cos(n-1)\theta - i\sin(n-1)\theta\}\,d\theta.$$

Now it is easily proved that

$$\int_\alpha^{\alpha+2\pi} \begin{matrix}\cos\\ \sin\end{matrix}\, m\theta\,d\theta = 0,$$

if m is an integer not zero.

Therefore $\displaystyle\int_{C'} \phi_r(z)\,dz = \int_\alpha^{\alpha+2\pi} b_{1,r}\,i\,d\theta = 2\pi i\,b_{1,r}.$

Therefore finally,

$$\int_C f(z)\,dz = \sum_{r=1}^{m}\int_C \phi_r(z)\,dz$$

$$= 2\pi i\sum_{r=1}^{m} b_{1,r}.$$

This result may be formally stated as follows:

If $f(z)$ be a function of z analytic at all points inside a closed contour C with the exception of a number of poles, and continuous throughout C and its interior (except at the poles), then $\int_C f(z)\,dz$ is equal to $2\pi i$ multiplied by the sum of the residues of $f(z)$ at its poles inside C.

This theorem renders it possible to evaluate a large number of definite integrals; examples of such integrals are given in the next Chapter.

25. In the case of a function given by a simple formula, it is usually possible to determine by inspection the poles of the function.

To calculate the residue of $f(z)$ at a pole $z = a$, the method generally employed is to expand $f(a+t)$ in a series of *ascending* powers of t (a process which is justifiable[5] for sufficiently small values of $|t|$), and the coefficient of t^{-1} in the expansion is the required residue. In the case of a pole of the first order, usually

[5] By applying Taylor's Theorem (see § 34) to $(z-a)^n f(z)$.

called a simple pole, it is generally shorter, in practice, to evaluate $\lim_{z \to a} (z-a)f(z)$ by the rules for determining limits; a consideration of the expression for $f(z)$ in the neighbourhood of a pole shews that the residue is this limit, provided that the limit exists.

26 Liouville's Theorem. *Let $f(z)$ be analytic for all values of z and let $|f(z)|<K$ where K is a constant. Then $f(z)$ is a constant.*

Let z, z' be any two points and let C be a contour such that z, z' are inside it; then by § 22

$$f(z')-f(z)=\frac{1}{2\pi i}\int_C \left\{\frac{1}{\zeta-z'}-\frac{1}{\zeta-z}\right\} f(\zeta)\, d\zeta;$$

take C to be a circle whose centre is z and whose radius is $\rho \geqslant 2|z'-z|$. On C write $\zeta = z+\rho e^{i\theta}$; since $|\zeta - z'| \geqslant \frac{1}{2}\rho$ when ζ is on C, it follows that

$$\begin{aligned}|f(z')-f(z)| &= \left|\frac{1}{2\pi}\int_C \frac{(z'-z)f(\zeta)}{(\zeta-z')(\zeta-z)}\,d\zeta\right| \\ &< \frac{1}{2\pi}\int_0^{2\pi}\frac{|z'-z|\,.\,K}{\frac{1}{2}\rho}\,d\theta \\ &< 2|z'-z|K\rho^{-1}.\end{aligned}$$

This is true for all values of $\rho \geqslant 2|z'-z|$.

Make $\rho \to \infty$, keeping z and z' fixed; then it is obvious that $f(z')-f(z)=0$; that is to say, $f(z)$ is constant.

CHAPTER VI

THE EVALUATION OF DEFINITE INTEGRALS

§ 27. A circular contour.—§ 28. Integrals of rational functions.—§ 29. Integrals of rational functions, continued.—§ 30. Jordan's lemma.—§ 31. Applications of Jordan's lemma.—§ 32. Other definite integrals.—§ 33. Examples of contour integrals.

27. If $f(x, y)$ be a rational function of x and y, the integral

$$\int_0^{2\pi} f(\cos\theta, \sin\theta)\, d\theta$$

can be evaluated in the following manner:

Let $z = \cos\theta + i\sin\theta$, so that $z^{-1} = \cos\theta - i\sin\theta$; then we have

$$\int_C f\left\{\frac{1}{2}(z+z^{-1}),\ \frac{1}{2i}(z-z^{-1})\right\}\frac{dz}{iz} = \int_0^{2\pi} f(\cos\theta, \sin\theta)\, d\theta,$$

wherein the contour of integration, C, is a unit circle with centre at $z=0$. If $f(\cos\theta, \sin\theta)$ is finite when $0 \leqslant \theta \leqslant 2\pi$, the integrand on the left-hand side is a function of z which is analytic on C; and also analytic inside C except at a finite number of poles. Consequently

$$\int_0^{2\pi} f(\cos\theta, \sin\theta)\, d\theta$$

is equal to $2\pi i$ times the sum of the residues of

$$i^{-1}z^{-1}f\left\{\frac{1}{2}(z+z^{-1}),\ \frac{1}{2i}(z-z^{-1})\right\}$$

at those of its poles which are inside the circle $|z| = 1$.

Example. $\displaystyle\int_0^{2\pi}\frac{d\theta}{a+b\cos\theta} = \frac{2\pi}{\sqrt{(a^2-b^2)}}$, *that sign being given to the radical which makes* $|a - \sqrt{(a^2-b^2)}| < |b|$, *it being supposed that* a/b *is* not *a real number such that*[1] $-1 \leqslant a/b \leqslant 1$.

[1] Apart from this restriction a, b may be any numbers real or complex.

Making the above substitution,

$$\int_0^{2\pi} \frac{d\theta}{a+b\cos\theta} = \frac{2}{i}\int_C \frac{dz}{bz^2+2az+b}$$
$$= \frac{2}{ib}\int_C \frac{dz}{(z-\alpha)(z-\beta)},$$

where α, β are the roots of $bz^2+2az+b=0$. The poles of the integrand are the points $z=\alpha$, $z=\beta$.

Since $|\alpha\beta|=1$, of the two numbers $|\alpha|$, $|\beta|$ one is greater than 1 and one less than 1, unless both are equal to 1. If both are equal to 1, put $\alpha=\cos\gamma+i\sin\gamma$, where γ is real; then $\beta=\alpha^{-1}=\cos\gamma-i\sin\gamma$, so that $2a/b=-\alpha-\beta=-2\cos\gamma$, i.e. $-1\leqslant a/b\leqslant 1$, which is contrary to hypothesis.

Let $\alpha=\dfrac{-a+\sqrt{(a^2-b^2)}}{b}$, $\beta=\dfrac{-a-\sqrt{(a^2-b^2)}}{b}$, that sign being given to the radical which makes $|a-\sqrt{(a^2-b^2)}|<|b|$; so that $|\alpha|<1$, $|\beta|>1$; then $z=\alpha$ is the only pole of the integrand inside C, and the residue of $\{(z-\alpha)(z-\beta)\}^{-1}$ at $z=\alpha$ is $(\alpha-\beta)^{-1}$.

Therefore
$$\int_0^{2\pi} \frac{d\theta}{a+b\cos\theta} = 2\pi i\times\frac{2}{ib}\times\frac{1}{\alpha-\beta}$$
$$= \frac{2\pi}{\sqrt{(a^2-b^2)}}.$$

28. *If $P(x)$, $Q(x)$ be polynomials in x such that $Q(x)$ has no real linear factors and the degree of $P(x)$ is less than the degree of $Q(x)$ by at least 2, then*[2] $\int_{-\infty}^{\infty}\frac{P(x)}{Q(x)}dx$ *is equal to $2\pi i$ times the sum of the residues of $P(z)/Q(z)$ at those of its poles which lie in the half plane above the real axis.*

Let
$$P(x)=a_0x^n+a_1x^{n-1}+\dots+a_n,$$
$$Q(x)=b_0x^m+b_1x^{m-1}+\dots+b_m,$$

where $m-n\geqslant 2$, $a_0\neq 0$, $b_0\neq 0$; choose r so large that

$$\frac{|a_1|}{r}+\frac{|a_2|}{r^2}+\dots+\frac{|a_n|}{r^n}\leqslant|a_0|,$$

and
$$\frac{|b_1|}{r}+\frac{|b_2|}{r^2}+\dots+\frac{|b_m|}{r^m}\leqslant\tfrac{1}{2}|b_0|;$$

[2] The reader will remember that an infinite integral is defined in the following manner: if $\int_a^R f(x)\,dx=g(R)$, then $\int_a^\infty f(x)\,dx$ means $\lim_{R\to\infty} g(R)$.

then if $|z| \geqslant r$, $$|z^{-n} P(z)| = \left| a_0 + \frac{a_1}{z} + \ldots + \frac{a_n}{z^n} \right|$$
$$\leqslant |a_0| + \frac{|a_1|}{r} + \ldots + \frac{|a_n|}{r^n}$$
$$\leqslant 2|a_0|,$$

and $$|z^{-m} Q(z)| \geqslant |b_0| - \left| \frac{b_1}{z} + \frac{b_2}{z^2} + \ldots + \frac{b_m}{z^m} \right|$$
$$\geqslant |b_0| - \frac{|b_1|}{r} - \frac{|b_2|}{r^2} - \ldots - \frac{|b_m|}{r^m}$$
$$\geqslant \tfrac{1}{2} |b_0|,$$

so that, if $|z| \geqslant r$, then $|z^{m-n} P(z)/Q(z)| \leqslant 4 |a_0 b_0^{-1}|$.

Let C be a contour consisting of that portion of the real axis which joins the points $-\rho$, $+\rho$ and of a semicircle, Γ, above the real axis, whose centre is the origin and whose radius is ρ; where ρ is any number greater than r.

Consider $\int_C \frac{P(z)}{Q(z)} dz$; this integral is equal to

$$\int_{-\rho}^{\rho} \frac{P(z)}{Q(z)} dz + \int_\Gamma \frac{P(z)}{Q(z)} dz.$$

Now $P(z)/Q(z)$ has no poles outside the circle $|z| = r$; for outside this circle $|P(z)/Q(z)| \leqslant 4 |a_0 b_0^{-1}| r^{n-m}$.

Therefore $$\int_{-\rho}^{\rho} \frac{P(z)}{Q(z)} dz + \int_\Gamma \frac{P(z)}{Q(z)} dz$$

is equal to $2\pi i$ times the sum of the residues of $P(z)/Q(z)$ at its poles inside C; i.e. at its poles in the half plane above the real axis.

Further, putting $z = \rho(\cos\theta + i \sin\theta)$ on Γ,

$$\left| \int_\Gamma \frac{P(z)}{Q(z)} dz \right| = \left| \int_0^\pi \frac{P(z)}{Q(z)} \rho(\cos\theta + i\sin\theta)\, i d\theta \right|$$
$$\leqslant \int_0^\pi \left| \frac{P(z)}{Q(z)} \right| \rho d\theta$$
$$\leqslant \int_0^\pi 4 |a_0 b_0^{-1} z^{n-m}| \rho d\theta$$
$$\leqslant 4\pi |a_0 b_0^{-1}| \rho^{n-m+1}.$$

Since $n - m + 1 \leqslant -1$, $\lim_{\rho \to \infty} \int_\Gamma \frac{P(z)}{Q(z)} dz = 0$.

But[3] $$\int_{-\infty}^{\infty} \frac{P(z)}{Q(z)} dz = \lim_{\rho, \sigma \to \infty} \int_{-\rho}^{\sigma} \frac{P(z)}{Q(z)} dz$$

$$= \lim_{\rho \to \infty} \int_{-\rho}^{\rho} \frac{P(z)}{Q(z)} dz$$

$$= - \lim_{\rho \to \infty} \int_{\Gamma} \frac{P(z)}{Q(z)} dz + 2\pi i \Sigma r_p,$$

where Σr_p means the sum of the residues of $P(z)/Q(z)$ at its poles in the half plane above the real axis.

Since it has been shewn that $\lim_{\rho \to \infty} \int_{\Gamma} \frac{P(z)}{Q(z)} dz = 0$, this is the theorem stated.

Example. *If* $a > 0$, $\int_{-\infty}^{\infty} \frac{dx}{(x^2 + a^2)^2} = \frac{\pi}{2a^3}$.

The value of the integral is $2\pi i$ times the residue of $(z^2+a^2)^{-2}$ at ai; putting $z = ai + t$,

$$\frac{1}{(z^2+a^2)^2} = \frac{1}{t^2(2ai+t)^2} = -\frac{1}{4a^2t^2} - \frac{i}{4a^3t} + \text{terms which are finite when } t=0.$$

The residue is therefore $-i/(4a^3)$; and hence the integral is equal to $\pi/(2a^3)$.

If $Q(x)$ has non-repeated real linear factors, the *principal value*[4] of the integral, which is written $P\int_{-\infty}^{\infty} \frac{P(x)}{Q(x)} dx$, exists. *Its value is* $2\pi i$ *times the sum of the residues of* $P(z)/Q(z)$ *at those of its poles which lie in the half plane above the real axis plus* πi *times the sum of the residues at those of the poles which lie on the real axis.*

To prove this theorem, let a_p' be a real root of $Q(x)$. Modify the contour by omitting the parts of the real axis between $a_p' - \delta_p$ and $a_p' + \delta_p$ and inserting a semicircle r_p, of radius δ_p and centre a_p, above the real axis; carry out this process for each real root. When a contour is modified in this way, so that its interior is diminished, the contour is said to be *indented.*

The limit of the integral along the surviving parts of the real axis when the numbers δ_p tend to zero is $P\int_{-\infty}^{\infty} \frac{P(x)}{Q(x)} dx$.

[3] Since $\lim_{\rho, \sigma \to \infty} \int_{-\rho}^{\sigma}$ *exists*, it is equal to $\lim_{\rho \to \infty} \int_{-\rho}^{\rho}$.

[4] Bromwich, *Theory of Infinite Series*, p. 415. The use of the letter P in two senses will not cause confusion.

If r_p' be the residue of $P(z)/Q(z)$ at a_p', the integral along the semicircle γ_p is $-\int_0^\pi \frac{P(z)}{Q(z)}(z-a_p')\,id\theta$, where $z=a_p'+\delta_p e^{i\theta}$, and this tends to $-\pi i r_p'$ as $\delta_p \to 0$; hence

$$P\int_{-\infty}^{\infty}\frac{P(x)}{Q(x)}dx-\pi i\Sigma r_p'=2\pi i\Sigma r_p.$$

29. A more interesting integral than the last is $\int_0^\infty \frac{P(x)}{Q(x)}dx$, where $P(x)$, $Q(x)$ are polynomials in x such that $Q(x)$ does not vanish for positive (or zero) values of x, and the degree of $P(x)$ is lower than that of $Q(x)$ by at least 2.

The value of this integral is the sum of the residues of

$$\log(-z)\,P(z)/Q(z)$$

at the zeros of $Q(z)$; *where the imaginary part of* $\log(-z)$ *lies between* $\pm i\pi$.

[The reader may obtain the formula for the principal value of the integral when $Q(x)=0$ has non-repeated positive roots.]

Consider $\int \log(-z)\frac{P(z)}{Q(z)}dz$, taken round a contour consisting of the arcs of circles of radii[5] R, δ, and the straight lines joining their end-points. On the first circle $-z=Re^{i\theta}$ $(-\pi\leqslant\theta\leqslant\pi)$; on the second circle $-z=\delta e^{i\theta}$ $(-\pi\leqslant\theta\leqslant\pi)$. And $\log(-z)$ is to be interpreted as $\log|z|+i\arg(-z)$, where $-\pi\leqslant\arg(-z)\leqslant\pi$; on one of the straight lines joining δ to R, $\arg(-z)=\pi$, on the other $\arg(-z)=-\pi$.

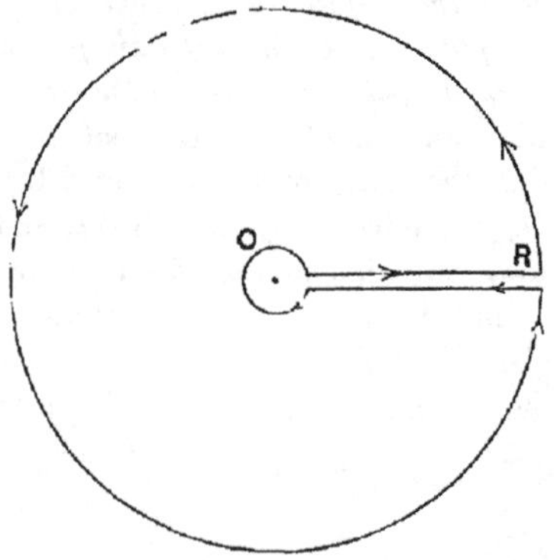

(The path of integration is not, strictly speaking, what has been previously defined as a contour, but the region bounded by the two arcs

[5] In future, the letter R will not be used to denote 'the real part of.'

and the straight lines is obviously one to which Goursat's lemma and the analysis of Chapter III can be applied.)

With the conventions as to $\log(-z)$, $\log(-z)\,P(z)/Q(z)$ *is analytic inside the contour except at the zeros of* $Q(z)$.

As in § 28 we can choose R_0 and δ_0 so that $|z^2 P(z)/Q(z)|$ does not exceed a fixed number, K, when $|z|=R>R_0$, and so that $|P(z)/Q(z)|<K$ when $|z|=\delta<\delta_0$; where K is independent of δ and R.

Let the circle of radius R be called Γ, and the circle of radius δ be called γ; let c_1, c_2 be the lines $\arg(-z)=-\pi$, $\arg(-z)=\pi$.

Then the integral round the contour is $2\pi i$ times the sum of the residues of $\log(-z)\,P(z)/Q(z)$ at the zeros of $Q(z)$ (these are the only poles of the integrand within the contour).

But the integral round the contour $=\int_\Gamma+\int_{c_2}+\int_\gamma+\int_{c_1}$.

Now
$$\left|\int_\Gamma\right| \leqslant \int_{-\pi}^{\pi}\left|\log(-z)\frac{P(z)}{Q(z)}\frac{dz}{d\theta}\right|d\theta$$
$$<\int_{-\pi}^{\pi}|(\log R+i\theta)|\,KR^{-1}\,d\theta,$$
which $\to 0$ as $R\to\infty$, since $R^{-1}\log R\to 0$ as $R\to\infty$.

So also
$$\left|\int_\gamma\right| \leqslant \int_{-\pi}^{\pi}\left|\log(-z)\frac{P(z)}{Q(z)}\frac{dz}{d\theta}\right|d\theta,\ \text{where } z=\delta e^{i\theta}$$
$$<\int_{-\pi}^{\pi}|\log\delta+i\theta|\,K\delta\,d\theta,$$
which tends to 0 as $\delta\to 0$, since $\delta\log\delta\to 0$ as $\delta\to 0$.

Put $-z=xe^{-i\pi}$ on c_1, $-z=xe^{i\pi}$ on c_2. Then
$$\int_{c_1}=\int_R^\delta(\log x-i\pi)\frac{P(x)}{Q(x)}dx$$
$$=\int_\delta^R(i\pi-\log x)\frac{P(x)}{Q(x)}dx,$$
and
$$\int_{c_2}=\int_\delta^R(i\pi+\log x)\frac{P(x)}{Q(x)}dx.$$

Hence $2\pi i$ times the sum of the residues of $\log(-z)\,P(z)/Q(z)$ at the zeros of $Q(z)$
$$=\lim_{\delta\to 0,\,R\to\infty}\left\{\int_\delta^R(i\pi-\log x)\frac{P(x)}{Q(x)}dx+\int_\delta^R(i\pi+\log x)\frac{P(x)}{Q(x)}dx\right\}$$
$$=\int_0^\infty 2\pi i\frac{P(x)}{Q(x)}dx,$$
which proves the proposition.

The interest of this integral lies in the fact that we apply Cauchy's theorem to a particular value (or *branch*[6]) of a many valued function.

If $P(x)$, $Q(x)$ be polynomials in x such that $Q(x)$ has no repeated positive roots and $Q(0) \neq 0$, and the degree of $P(x)$ is less than that of $Q(x)$ by at least 1, the reader may prove, by integrating the branch of $(-z)^{a-1} P(z)/Q(z)$ for which $|\arg(-z)| \leqslant \pi$ round the contour of the preceding example and proceeding to the limit when $\delta \to 0$, $R \to \infty$, that, if $0 < a < 1$ and x^{a-1} means the positive value of x^{a-1}, then

$$P\int_0^\infty x^{a-1}\frac{P(x)}{Q(x)}\,dx = \pi \operatorname{cosec}(a\pi)\,\Sigma r - \pi\cot(a\pi)\,\Sigma r',$$

where Σr means the sum of the residues of $(-z)^{a-1} P(z)/Q(z)$ at those zeros of $Q(z)$ at which z is not positive, and $\Sigma r'$ means the sum of the residues of $x^{a-1} P(x)/Q(x)$ at those zeros of $Q(x)$ at which x is positive. When $Q(x)$ has zeros at which x is positive, the lines c_1, c_2 have to be indented as in the last example of § 28.

Examples. *If* $0<a<1$,

$$\int_0^\infty \frac{x^{a-1}}{1+x}\,dx = \frac{\pi}{\sin a\pi}, \quad P\int_0^\infty \frac{x^{a-1}}{1-x}\,dx = \pi\cot a\pi.$$

30. In connection with examples of the type which will next be considered, the following lemma is frequently useful.

JORDAN'S LEMMA[7]. *Let $f(z)$ be a function of z which satisfies the following conditions when $|z| > c$ and the imaginary part of z is not negative (c being a positive constant):*

(i) *$f(z)$ is analytic,*

(ii) *$|f(z)| \to 0$ uniformly as $|z| \to \infty$.*

Let m be a positive constant, and let Γ be a semicircle of radius R ($> c$), above the real axis, and having its centre at the origin.

Then $$\lim_{R\to\infty}\left(\int_\Gamma e^{miz} f(z)\,dz\right) = 0.$$

If we put $z = R(\cos\theta + i\sin\theta)$, θ increases from 0 to π as z describes Γ.

Therefore $$\int_\Gamma e^{miz} f(z)\,dz = \int_0^\pi z e^{miz} f(z)\,i\,d\theta,$$

[6] Forsyth, *Theory of Functions*, Chapter VIII.

[7] Jordan, *Cours d'Analyse*, t. II, § 270.

so that
$$\left|\int_\Gamma e^{miz} f(z)\,dz\right| = \left|\int_0^\pi z e^{miz} f(z)\, i\,d\theta\right|$$
$$\leqslant \int_0^\pi |ize^{miz} f(z)|\,d\theta$$
$$\leqslant \int_0^\pi \eta R e^{-mR\sin\theta}\,d\theta,$$
where η is the greatest value of $|f(z)|$ when $|z| = R$.

In the last integral, divide the range of integration into two parts, viz. from 0 to $\frac{1}{2}\pi$ and from $\frac{1}{2}\pi$ to π; write $\pi-\theta$ for θ in the second part; then, noting that, when[8] $0 \leqslant \theta \leqslant \frac{1}{2}\pi$, $\sin\theta \geqslant 2\theta/\pi$, so that $e^{-mR\sin\theta} \leqslant e^{-2mR\theta/\pi}$, we see that
$$\left|\int_\Gamma e^{miz} f(z)\, z^{-1}\,dz\right| \leqslant 2\eta R \int_0^{\frac{1}{2}\pi} e^{-mR\sin\theta}\,d\theta$$
$$\leqslant 2\eta R \int_0^{\frac{1}{2}\pi} e^{-2mR\theta/\pi}\,d\theta$$
$$\leqslant \frac{2\eta\pi}{2m}(1-e^{-mR}) < \eta\pi m^{-1}.$$

But $\eta \to 0$ as $R \to \infty$; and therefore
$$\lim_{R\to\infty}\left|\int_\Gamma e^{miz} f(z)\,dz\right| = 0.$$

31. The following theorem may be proved, by the use of Jordan's lemma:

Let $P(x)$, $Q(x)$ be polynomials in x such that $Q(x)$ has no real linear factors, and the degree of $P(x)$ does not exceed the degree of $Q(x)$, and let $m > 0$.

Then
$$\int_0^\infty \left\{\frac{P(x)}{Q(x)} e^{mix} - \frac{P(-x)}{Q(-x)} e^{-mix}\right\}\frac{dx}{x}$$
is equal to $\pi i P(0)/Q(0)$ plus $2\pi i$ times the sum of the residues of $\dfrac{P(z)}{Q(z)}\dfrac{e^{miz}}{z}$ at the zeros of $Q(z)$ in the half plane above the real axis.

[The reader may obtain the formula for the principal value of the integral when $Q(x)=0$ has non-repeated real roots.]

Consider $\displaystyle\int_C \frac{P(z)}{Q(z)} e^{miz}\frac{dz}{z}$ taken along a contour C consisting of the straight line joining $-R$ to $-\delta$, a semicircle, γ, of radius δ, above the real axis, and with its centre at the origin, the straight line joining δ to

[8] If we draw the graphs $y=\sin x$, $y=2x/\pi$, this inequality appears obvious; it may be proved by shewing that $\dfrac{\sin\theta}{\theta}$ decreases as θ increases from 0 to $\frac{1}{2}\pi$.

R, and the semicircle Γ, where $\delta < \delta_0$, $R > R_0$, and δ and R are to be so chosen that all the poles of $P(z)/Q(z)$ lie outside the circle $|z| = \delta_0$ and inside the circle $|z| = R_0$.

Then $\int_C \frac{P(z)}{Q(z)} e^{miz} \frac{dz}{z}$ is equal to $2\pi i$ times the sum of the residues of $\frac{P(z)}{Q(z)} \frac{e^{miz}}{z}$ at the zeros of $Q(z)$ which lie in the upper half plane, as may be shewn by analysis similar to that of § 28.

But $$\int_C \frac{P(z)}{Q(z)} e^{miz} \frac{dz}{z} = \left(\int_{-R}^{-\delta} + \int_\gamma + \int_\delta^R + \int_\Gamma\right) \frac{P(z)}{Q(z)} e^{miz} \frac{dz}{z}.$$

Put $z = -x$ in the first integrand, and $z = x$ in the third; then

$$\left(\int_{-R}^{-\delta} + \int_\delta^R\right) \frac{P(z)}{Q(z)} e^{miz} \frac{dz}{z} = \int_\delta^R \left\{\frac{P(x)}{Q(x)} e^{mix} - \frac{P(-x)}{Q(-x)} e^{-mix}\right\} \frac{dx}{x}.$$

Since $\lim\limits_{|z|\to\infty} P(z)/Q(z)$ is finite, $\int_\Gamma \frac{P(z)}{Q(z)} e^{miz} \frac{dz}{z} \to 0$ as $R \to \infty$, by Jordan's lemma.

Also, putting $z = \delta e^{i\theta}$ on γ,

$$\int_\gamma \frac{P(z)}{Q(z)} e^{miz} \frac{dz}{z} = -\int_0^\pi \left\{\frac{P(0)}{Q(0)} + \delta e^{i\theta} f(z)\right\} e^{miz}\, i\, d\theta,$$

where $|f(z)|$ does not exceed a number independent of z when $\delta < \delta_0$.

Hence $$\lim_{\delta\to 0} \int_\gamma \frac{P(z)}{Q(z)} e^{miz} \frac{dz}{z} = -\pi i \frac{P(0)}{Q(0)}.$$

Making $R \to \infty$, $\delta \to 0$ in the above formula for $\int_C \frac{P(z)}{Q(z)} e^{miz} \frac{dz}{z}$, the result stated follows at once.

Corollary. Put $P(z) = Q(z) = 1$; then, if $m > 0$,

$$\int_0^\infty \frac{\sin mx}{x} dx = \tfrac{1}{2}\pi.$$

By arguments similar to those used in proving Jordan's lemma, we may shew that

$$\int_0^\infty e^{a \cos bx} \sin(a \sin bx) \frac{x\,dx}{x^2 + r^2} = \tfrac{1}{2}\pi (e^{ae^{-br}} - 1),$$

if $$a > 0,\ b > 0,\ r > 0.$$

Consider $\int_C e^{ae^{biz}} \frac{z\,dz}{z^2 + r^2}$, where the contour C consists of the straight line joining the points $-R$, R and of the semicircle Γ, where $R > 2r$.

The only pole of the integrand inside the contour is at $z = ri$; and the residue of the integrand is easily found to be $\frac{1}{2} e^{ae^{-br}}$.

Therefore $$\left(\int_{-R}^{0} + \int_{0}^{R} + \int_{\Gamma}\right) e^{ae^{bzi}} \frac{z\,dz}{z^2+r^2} = \pi i e^{ae^{-br}}.$$

In the first integral put $z=-x$, in the second put $z=x$.

Then
$$2i\int_0^R e^{a\cos bx}\sin(a\sin bx)\frac{x\,dx}{x^2+r^2} = \pi i e^{ae^{-br}} - \int_\Gamma e^{ae^{bzi}}\frac{z\,dz}{z^2+r^2}.$$

Now
$$e^{ae^{bzi}} = 1 + ae^{bzi} + \frac{1}{2!}a^2e^{2bzi} + \dots$$
$$= 1 + ae^{bzi}w,$$

where
$$|w| = \left|1 + \frac{ae^{bzi}}{2!} + \frac{a^2e^{2bzi}}{3!} + \dots\right|$$
$$\leqslant 1 + \left|\frac{ae^{bzi}}{2!}\right| + \left|\frac{a^2e^{2bzi}}{3!}\right| + \dots,$$

and on Γ, $|e^{bzi}| = e^{-bR\sin\theta} \leqslant 1$, where $z = R(\cos\theta + i\sin\theta)$; so that
$$|w| < 1 + a + \frac{a^2}{2!} + \dots$$
$$= e^a.$$

Consequently, putting $z = R(\cos\theta + i\sin\theta)$ on Γ,
$$\int_\Gamma e^{ae^{bzi}}\frac{z\,dz}{z^2+r^2} = \int_0^\pi (1 + ae^{bzi}w)\left(1 - \frac{r^2}{z^2+r^2}\right) i\,d\theta$$
$$= \pi i - r^2 i\int_0^\pi \frac{d\theta}{z^2+r^2} - \int_0^\pi ae^{bzi}w\frac{z^2}{z^2+r^2}\,i\,d\theta.$$

But
$$|z^2+r^2| \geqslant |z|^2 - r^2 > \tfrac{3}{4}R^2,$$

and therefore
$$\left|\int_0^\pi \frac{d\theta}{z^2+r^2}\right| < \frac{4\pi}{3R^2},$$

while
$$\left|\int_0^\pi ae^{bzi}w\frac{z^2}{z^2+r^2}\,i\,d\theta\right| \leqslant \int_0^\pi \left|ae^{bzi}w\frac{z^2}{z^2+r^2}\,i\,d\theta\right|$$
$$< \int_0^\pi \tfrac{4}{3}ae^{-bR\sin\theta}e^a\,d\theta$$
$$< \frac{4\pi ae^a}{3bR}, \text{ as in Jordan's lemma.}$$

Consequently
$$\int_\Gamma e^{ae^{bzi}}\frac{z\,dz}{z^2+r^2} = \pi i + \epsilon_R,$$

where $\lim\limits_{R\to\infty}\epsilon_R = 0$; and therefore, by the definition of an infinite integral,
$$2i\int_0^\infty e^{a\cos bx}\sin(a\sin bx)\frac{x\,dx}{x^2+r^2} = \pi i e^{ae^{-br}} - \pi i.$$

32. Infinite integrals involving hyperbolic functions can frequently be evaluated by means of a contour in the shape of a rectangle; an example of such an integral is the following:

$$\int_0^\infty \frac{\cosh ax}{\cosh \pi x}\,dx = \tfrac{1}{2}\sec \tfrac{1}{2}a, \textit{ when } -\pi < a < \pi.$$

Consider $\int_\Gamma \frac{e^{az}}{\cosh \pi z}\,dz$ taken along the contour Γ formed by the rectangle whose corners have complex coordinates $-R$, R, $R+i$, $-R+i$, where $R>0$; let these corners be A, B, C, D. The zeros of

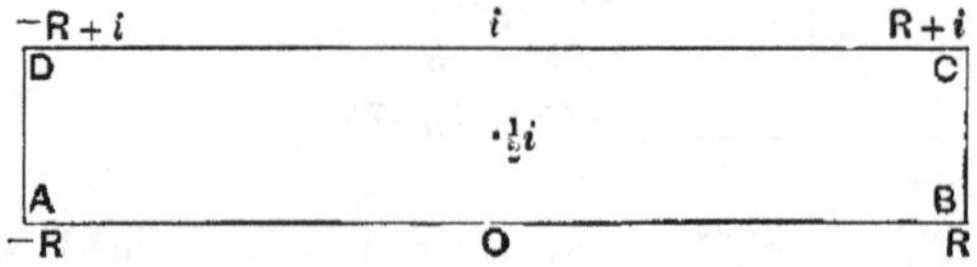

$\cosh \pi z$ are at the points $z=(n+\frac{1}{2})\,i$, where n is any integer; so that the only pole of the integrand inside the contour is at the point $z=\frac{1}{2}i$. If $z=\frac{1}{2}i+t$, then

$$\frac{e^{az}}{\cosh \pi z} = \frac{e^{\frac{1}{2}ai}\,(1+at+\frac{1}{2}a^2t^2+\dots)}{i\sinh \pi t}$$

$$= \frac{e^{\frac{1}{2}ai}}{\pi i t}\,\frac{(1+at+\frac{1}{2}a^2t^2+\dots)}{(1+\frac{1}{6}\pi^2t^2+\dots)},$$

so that the residue of $e^{az}\operatorname{sech}\pi z$ at $\frac{1}{2}i$ is $e^{\frac{1}{2}ai}/(\pi i)$.

Therefore $$\int_\Gamma \frac{e^{az}}{\cosh \pi z}\,dz = 2e^{\frac{1}{2}ai}.$$

Now $$\int_\Gamma \frac{e^{az}}{\cosh \pi z}\,dz = \left(\int_{AB}+\int_{BC}+\int_{CD}+\int_{DA}\right)\frac{e^{az}}{\cosh \pi z}\,dz;$$

on AB we may put $z=x$ where x is real; on CD we may put $z=i+x$ where x is real; on BC we may put $z=R+iy$ where y is real; and on DA we may put $z=-R+iy$ where y is real. Therefore

$$\int_\Gamma \frac{e^{az}}{\cosh \pi z}\,dz = \int_{-R}^{R}\frac{e^{ax}}{\cosh \pi x}\,dx + \int_R^{-R}\frac{e^{a(i+x)}}{\cosh \pi (i+x)}\,dx$$

$$+\int_0^1 \frac{e^{a(R+iy)}}{\cosh \pi (R+iy)}\,i\,dy + \int_1^0 \frac{e^{-a(R-iy)}}{\cosh \pi (R-iy)}\,i\,dy$$

$$=(1+e^{ai})\int_{-R}^{R}\frac{e^{ax}}{\cosh \pi x}\,dx + \epsilon_R,$$

where $$\epsilon_R = \int_0^1 \frac{e^{a(R+iy)}}{\cosh \pi (R+iy)} i\,dy - \int_0^1 \frac{e^{-a(R-iy)}}{\cosh \pi (R-iy)} i\,dy;$$

also $$\int_{-R}^{R} \frac{e^{ax}}{\cosh \pi x} dx = \int_0^R \frac{e^{ax}}{\cosh \pi x} dx + \int_{-R}^{0} \frac{e^{ax}}{\cosh \pi x} dx$$

$$= \int_0^R \frac{e^{ax}}{\cosh \pi x} dx + \int_0^R \frac{e^{-ax}}{\cosh \pi x} dx,$$

on writing $-x$ for x in the second integral.

Therefore $$2e^{\frac{1}{2}ai} = 2(1+e^{ai}) \int_0^R \frac{\cosh ax}{\cosh \pi x} dx + \epsilon_R.$$

Now $$|\epsilon_R| \leqslant \left|\int_0^1 \frac{e^{a(R+iy)}}{\cosh \pi (R+iy)} i\,dy\right| + \left|\int_0^1 \frac{e^{-a(R-iy)}}{\cosh \pi (R-iy)} i\,dy\right|$$

$$\leqslant \int_0^1 \frac{|e^{a(R+iy)}|\,dy}{|\cosh \pi (R+iy)|} + \int_0^1 \frac{|e^{-a(R-iy)}|\,dy}{|\cosh \pi (R-iy)|}.$$

Also $$|2\cosh \pi (R \pm iy)| = |e^{\pi(R\pm iy)} + e^{-\pi(R\pm iy)}|$$

$$\geqslant |e^{\pi(R\pm iy)}| - |e^{-\pi(R\pm iy)}|$$

$$\geqslant e^{\pi R} - e^{-\pi R}.$$

Therefore $$|\epsilon_R| \leqslant \int_0^1 \frac{e^{aR}}{\sinh \pi R} dy + \int_0^1 \frac{e^{-aR}}{\sinh \pi R} dy$$

$$\leqslant 2\frac{\cosh (aR)}{\sinh (\pi R)}.$$

But, if $-\pi < a < \pi$, $\lim_{R\to\infty} 2\frac{\cosh (aR)}{\sinh (\pi R)} = 0$; therefore, if $-\pi < a < \pi$, $\lim_{R\to\infty} \epsilon_R = 0$.

But $\int_0^R \frac{\cosh ax}{\cosh \pi x} dx$ is equal to $\frac{2e^{\frac{1}{2}ai} - \epsilon_R}{2(1+e^{ai})}$; and therefore

$$\int_0^\infty \frac{\cosh ax}{\cosh \pi x} dx = \tfrac{1}{2}\sec \tfrac{1}{2}a, \quad (-\pi < a < \pi).$$

33. Solutions of the following examples, of which the earlier ones are taken from recent College and University Examination Papers, can be obtained by the methods developed in this chapter.

1. Shew that $$\int_0^{2\pi} \frac{\sin^2\theta}{a + b\cos\theta} d\theta = \frac{2\pi}{b^2}\left(a - \sqrt{a^2-b^2}\right),$$

when $a > b > 0$. Give reasons why this equation should still be true when $a = b$. (Math. Trip. 1904.)

2. Evaluate $\int_0^\infty \frac{dx}{(x^2+f^2)^2(x^2+g^2)^2}$ when $f>0$, $g>0$.

(Trinity, 1905.)

3. Shew that $\int_{-\infty}^\infty \frac{dx}{(x^2+b^2)(x^2+a^2)^2} = \frac{\pi(2a+b)}{2a^3b(a+b)^2}$ when $a>0$, $b>0$.

(Whittaker, *Modern Analysis*.)

4. Prove that, if $a>0$, $b>0$, $c>0$, $b^2-ac>0$, then

$$\int_0^\infty \frac{dx}{a^2+2b^2x^2+c^2x^4} = \frac{\pi}{2^{\frac{3}{2}}a\sqrt{(b^2+ac)}}.$$ (Trinity, 1908.)

5. Evaluate $\int_0^\infty \frac{dx}{(x^2+a^2)^2(x^2+b^2)^2(x^2+c^2)^2}$ when a, b, c are real.

(St John's, 1907.)

6. Shew that, if $a>0$,

$$\int_0^\infty \frac{x^2\,dx}{(a^4+x^4)^3}\,dx = \frac{5\pi\sqrt{2}}{128a^9}.$$ (Trinity, 1902.)

7. Shew that $\int_0^\infty \frac{\sinh ax}{\sinh \pi x}\,dx = \frac{1}{2}\tan\frac{1}{2}a$ when $-\pi<a<\pi$.

8. Shew that $$\int_0^\infty \frac{x^2}{\sinh^2 x}\,dx = \frac{\pi^2}{6}.$$ (Clare, 1903.)

9. By integrating $\int e^{-z^2}dz$ round a rectangle, shew that

$$\int_{-\infty}^\infty e^{-t^2}\cos 2at\,.\,dt = e^{-a^2}\Gamma(\tfrac{1}{2}), \qquad \int_{-\infty}^\infty e^{-t^2}\sin 2at\,.\,dt = 0.$$

10. Shew that $$\int_0^\infty \frac{\sin x}{\sinh x}\,dx = \frac{1}{2}\pi\tanh\frac{1}{2}\pi.$$ (Clare, 1905.)

11. Shew that $\int_0^\infty \frac{x\cos ax}{\sinh x}\,dx = \frac{\pi^2 e^{-a\pi}}{(1+e^{-a\pi})^2}$, when a is real.

(Math. Trip. 1906.)

12. Evaluate $\int \frac{dz}{1+z^4}$ taken round the ellipse whose equation is $x^2-xy+y^2+x+y=0$. (Clare, 1903.)

13. Shew that, if $m>0$, $a>0$,

$$\int_0^\infty \frac{x\sin mx}{a^4+x^4}\,dx = \frac{\pi}{2a^2}e^{-\frac{ma}{\sqrt{2}}}\sin\frac{ma}{\sqrt{2}}.$$ (Trinity, 1906.)

14. Shew that $\int_0^\infty \frac{\sin^3 x \cos x}{x^3} dx = \frac{1}{4}\pi$. (Peterhouse, 1905.)

15. Shew that, if $m \geqslant 0$, $a > 0$,

$$\int_0^\infty \frac{\cos mx \,.\, dx}{a^2 + x^2} = \frac{\pi}{2a} e^{-ma}, \quad \int_0^\infty \frac{\cos mx \,.\, dx}{(1 + x^2)^2} = \frac{\pi}{4} e^{-m} (1 + m).$$

(Peterhouse, 1907.)

16. Shew that $\int_0^\infty \frac{\cos^2 x}{(1+x^2)^2} dx = \frac{\pi}{8}(1 + 3e^{-2})$.

(Peterhouse, 1907.)

17. Shew that, if $a > 0$,

$$\int_0^\infty \frac{x - \sin x}{x^3 (a^2 + x^2)} dx = \tfrac{1}{2}\pi a^{-4} (\tfrac{1}{2} a^2 - a + 1 - e^{-a}).$$

(Math. Trip. 1902.)

18. Shew that, if $m > 0$, $a > 0$,

$$\int_0^\infty \frac{\cos mx}{a^4 + x^4} dx = \frac{\pi}{2a^3} e^{-ma/\sqrt{2}} \sin\left(\frac{ma}{\sqrt{2}} + \frac{\pi}{4}\right).$$

(Trinity, 1907.)

19. Shew that, if $m > 0$, $a > 0$,

$$\int_0^\infty \frac{\sin^2 mx}{x^2 (a^2 + x^2)} dx = \frac{\pi}{4a^3} (e^{-2ma} - 1 + 2ma).$$

(Trinity, 1912.)

20. Shew that, if $a > 0$,

$$\int_0^\infty \frac{\cos ax}{1 + x^2 + x^4} dx = \frac{\pi}{\sqrt{3}} \sin\left(\frac{a}{2} + \frac{\pi}{6}\right) e^{-\frac{a\sqrt{3}}{2}}.$$

(Clare, 1902.)

21. Shew that $\int_0^\infty \frac{\cos x - x \sin x}{1 + x^2} dx = 0$. (Trinity, 1903.)

22. Shew that, when n is an even positive integer,

$$\int_0^\infty \frac{1}{x^2+1} \frac{\sin nx}{\sin x} dx = \pi \frac{e^n - 1}{(e^2 - 1) e^{n-1}}.$$ (Jesus, 1903.)

23. By taking a quadrant of a circle indented at ai as contour, shew that, if $m > 0$, $a > 0$, then

$$\int_0^\infty \frac{x \cos mx \pm a \sin mx}{x^2 + a^2} dx = -e^{\pm ma} li (e^{\mp ma}).$$ (Schlomilch.)

[$li (e^{\pm x})$ is defined by Bromwich, *Infinite Series*, p. 325.]

24. Shew that, if $m>0$, $c>0$ and a is real,

$$\int_{-\infty}^{\infty} \frac{\sin m(x-a)}{x-a} \frac{dx}{x^2+c^2} = \frac{\pi}{a^2+c^2}\left\{1 - \frac{e^{-mc}}{c}(c\cos ma - a \sin ma)\right\}.$$

(Trinity, 1911.)

25. Shew that, if $m>n>0$ and a, b are real, then

$$\int_{-\infty}^{\infty} \frac{\sin m(x-a)}{x-a} \frac{\sin n(x-b)}{x-b}\, dx = \pi \frac{\sin n(a-b)}{a-b}.$$

(Math. Trip. 1909.)

26. Shew that, if $0<a<2$, then

$$\int_0^{\infty} \frac{\sin^2 x \sin ax}{x^2}\, dx = \tfrac{1}{2}\pi a - \tfrac{1}{8}\pi a^2. \qquad \text{(Legendre.)}$$

27. Shew that $$P\int_0^{\infty} \frac{\tan x}{x}\, dx = \tfrac{1}{2}\pi. \qquad \text{(Legendre.)}$$

28. By using the contour of § 29, shew that, if $-1<p<1$ and $-\pi<\lambda<\pi$, then

$$\int_0^{\infty} \frac{x^{-p}\, dx}{1+2x\cos\lambda + x^2} = \frac{\pi}{\sin p\pi} \frac{\sin p\lambda}{\sin \lambda}. \qquad \text{(Euler.)}$$

29. Draw the straight line joining the points $\pm i$, and the semi-circle of $|z|=1$ which lies on the right of this line. Let C be the ontour formed by indenting this figure at $-i$, 0, i. By considering $\int_C z^{n-1}(z+z^{-1})^m\, dz$, shew that, if $n>m>-1$, then

$$\int_{-\frac{1}{2}\pi}^{\frac{1}{2}\pi} e^{ni\theta}\cos^m\theta\, d\theta = 2^{1-m}\sin\tfrac{1}{2}(n-m)\pi \int_0^1 t^{n-m-1}(1-t^2)^m\, dt.$$

Deduce that[9]

$$\int_0^{\frac{1}{2}\pi} \cos n\theta \cos^m\theta\, d\theta = \frac{\pi\,\Gamma(m+1)}{2^{m+1}\Gamma(\frac{1}{2}m+\frac{1}{2}n+1)\Gamma(\frac{1}{2}m-\frac{1}{2}n+1)};$$

and from the formula $\cos(n+1)\theta + \cos(n-1)\theta = 2\cos\theta\cos n\theta$, establish this result for all real values of n if $m>-1$.

30. By integrating $\int e^{-z^2}dz$ round a rectangle whose corners are 0, R, $R+ai$, ai, shew that

$$\int_0^{\infty} e^{-t^2}\sin 2at\,.\,dt = e^{-a^2}\int_0^a e^{y^2}\, dy.$$

[9] The result $\Gamma(a)\Gamma(1-a) = \pi \operatorname{cosec} a\pi$, which is required in this example, may be established by writing $x=t/(1-t)$ in the first example at the end of § 29, when $0<a<1$, and making use of the value of the first Eulerian integral; it may be proved for all values of a by a use of the recurrence formula $\Gamma(a+1)=a\Gamma(a)$.

31. Let $Q(z)$ be a polynomial and let the real part of a be numerically less than π. By integrating $\int \frac{Q(z-\pi i)}{\cosh z + \cos a} dz$ round a rectangle, shew that

$$\int_{-\infty}^{\infty} \frac{Q(x+\pi i) - Q(x-\pi i)}{\cosh x + \cos a} dx = 2\pi \operatorname{cosec} a \{Q(ai) - Q(-ai)\}.$$

Deduce that

$$\int_0^{\infty} \frac{x^4\, dx}{\cosh x + \cos a} = \tfrac{1}{15} a (\pi^2 - a^2)(7\pi^2 - 3a^2) \operatorname{cosec} a.$$

32. Let Γ_1 be a contour consisting of the part of the real axis joining the points $\pm R$, and of a semicircle of radius R above the real axis, the contour being indented at the points $n\pi/b$ where n takes all integral values and $b > 0$; also let bR/π be half an odd integer. Let Γ_2 be the reflexion of Γ_1 in the real axis, properly oriented.

Shew that, if $-b < a < b$ and if $P(z)$, $Q(z)$ are polynomials such that $Q(z)$ has no real factors and the degree of $Q(z)$ exceeds that of $P(z)$, then

$$P\int_{-\infty}^{\infty} \frac{e^{aix}}{\sin bx} \frac{P(x)}{Q(x)} dx = \tfrac{1}{2} \lim \left\{ \int_{\Gamma_1} \frac{e^{aiz}}{\sin bz} \cdot \frac{P(z)}{Q(z)} dz - \int_{\Gamma_2} \frac{e^{aiz}}{\sin bz} \frac{P(z)}{Q(z)} dz \right\};$$

where the limit is taken by making $R \to \infty$ and the radii of the indentations tend to zero.

Deduce that $P\int_{-\infty}^{\infty} \frac{e^{aix}}{\sin bx} \frac{P(x)}{Q(x)} dx = \pi i (\Sigma r - \Sigma r')$,

where Σr means the sum of the residues of the integrand at its poles in the upper half-plane and $\Sigma r'$ the sum of the residues at the poles in the lower half-plane.

33. Shew that, if $-b < a < b$, then

$$P\int_0^{\infty} \frac{\sin ax}{\sin bx} \frac{dx}{1+x^2} = \tfrac{1}{2}\pi \frac{\sinh a}{\sinh b}, \quad P\int_0^{\infty} \frac{\cos ax}{\sin bx} \frac{x\, dx}{1+x^2} = \tfrac{1}{2}\pi \frac{\cosh a}{\sinh b},$$

$$P\int_0^{\infty} \frac{\sin ax}{\cos bx} \frac{dx}{x(1+x^2)} = \tfrac{1}{2}\pi \frac{\sinh a}{\cosh b}, \quad P\int_0^{\infty} \frac{\cos ax}{\cos bx} \frac{dx}{1+x^2} = \tfrac{1}{2}\pi \frac{\cosh a}{\cosh b}.$$

(Legendre, Cauchy.)

34. If $(2m-1)b < a < (2m+1)b$ and m is a positive integer, deduce from Example 33 that

$$P\int_0^\infty \frac{\sin ax}{\sin bx}\frac{dx}{1+x^2} = \tfrac{1}{2}\pi\frac{\cosh(a-2mb)-e^{-a}}{\sinh b},$$

and three similar results. (Legendre.)

35. Shew that $\displaystyle\int_0^\infty \frac{dx}{(1+x^2)\cosh(\frac{1}{2}\pi x)} = \log 2.$

(Math. Trip. 1906.)

[Take the contour of integration to be the square whose corners are $\pm N$, $\pm N + 2Ni$, where N is an integer; and make $N \to \infty$.]

The results of Examples 36—39, which are due to Hardy, may be obtained by integrating expressions of the type

$$\int \frac{e^{az}}{1+2pe^z \pm e^{2z}} \cdot \frac{dz}{z+ia}$$

round a contour similar to that of Example 35. In all the examples a and δ are real; and, in Examples 36 and 38, $-\pi < \delta < \pi$.

36. $$\int_0^\infty \frac{1}{\cosh x + \cos\delta}\frac{dx}{a^2+x^2} = \frac{2\delta\pi}{a\sin\delta}\sum_{n=0}^\infty \frac{1}{\{(2n+1)\pi+a\}^2-\delta^2}.$$

Deduce that

$$\int_0^\infty \frac{1}{\cosh x + \cos\delta}\frac{dx}{\pi^2+x^2} = \frac{1}{\delta\sin\delta} - \frac{1}{4\sin^2\frac{1}{2}\delta}.$$

37. $$\int_0^\infty \frac{1}{\cosh x + \cosh\delta}\frac{dx}{a^2+x^2} = \frac{2\delta\pi}{\sinh\delta}\sum_{n=0}^\infty \frac{1}{\{(2n+1)\pi+a\}^2+\delta^2},$$

$$\int_0^\infty \frac{\cosh\frac{1}{2}x}{\cosh x + \cosh\delta}\frac{dx}{a^2+x^2} = \frac{\pi}{a\cosh\frac{1}{2}\delta}\sum_{n=0}^\infty \frac{(-)^n\{(2n+1)\pi+a\}}{\{(2n+1)\pi+a\}^2+\delta^2}.$$

38. $$\int_0^\infty \frac{\cosh\frac{1}{2}x}{\cosh x + \cos\delta}\frac{dx}{a^2+x^2} = \frac{\pi}{a\cos\frac{1}{2}\delta}\sum_{n=0}^\infty \frac{(-)^n\{(2n+1)\pi+a\}}{\{(2n+1)\pi+a\}^2-\delta^2}.$$

Deduce that

$$\int_0^\infty \frac{dx}{\cosh(\frac{1}{2}x)\,.\,(a^2+x^2)} = \frac{2}{a}\int_0^1 \frac{t^{a/\pi}}{1+t^2}\,dt.$$

39. $$P\int_{-\infty}^\infty \frac{1}{\sinh x - \sinh\delta}\frac{dx}{\pi^2+x^2} = \frac{1}{\cosh\delta}\left\{\frac{\delta}{\delta^2+\pi^2}+\frac{1}{\delta}\right\} - \frac{1}{\sinh\delta}.$$

40. Shew that, if $a > 0$, $m > 0$, $-1 < r < 1$, then

$$\int_0^\infty \frac{x\,dx}{m^2+x^2}\,\frac{\sin 2ax}{1-2r\cos 2ax+r^2} = \frac{\frac{1}{2}\pi}{e^{2am}-r},$$

$$\int_0^\infty \frac{x\,dx}{m^2+x^2}\,\frac{\sin ax}{1-2r\cos 2ax+r^2} = \frac{\frac{1}{2}\pi\, e^{am}}{(1+r)(e^{2am}-r)}.$$

Shew that, if the principal values of the integrals are taken, the results are true when $r = 1$. (Legendre.)

41. By integrating $\int \frac{e^{\pm aiz}}{e^{2\pi z}-1}\,dz$ round the rectangle whose corners are 0, R, $R+i$, i, (the rectangle being indented at 0 and i) shew that, if a be real, then

$$\int_0^\infty \frac{\sin ax}{e^{2\pi x}-1}\,dx = \tfrac{1}{4}\coth\left(\tfrac{1}{2}a\right) - \tfrac{1}{2}a^{-1}. \qquad \text{(Legendre.)}$$

42. By employing a rectangle indented at $\frac{1}{2}i$, shew that, if a be real, then

$$\int_0^\infty \frac{\sin ax}{e^{2\pi x}+1}\,dx = \tfrac{1}{2}a^{-1} - \tfrac{1}{4}\operatorname{cosech}\left(\tfrac{1}{2}a\right). \qquad \text{(Legendre.)}$$

43. By integrating $\int e^{-\lambda z} z^{n-1}\,dz$ round the sector of radius R bounded by the lines $\arg z = 0$, $\arg z = a < \frac{1}{2}\pi$, (the sector being indented at 0), shew that, if $\lambda > 0$, $n > 0$, then

$$\int_0^\infty x^{n-1} e^{-\lambda x\cos a}\cos(\lambda x\sin a)\,dx = \lambda^{-n}\Gamma(n)\cos na,$$

$$\int_0^\infty x^{n-1} e^{-\lambda x\cos a}\sin(\lambda x\sin a)\,dx = \lambda^{-n}\Gamma(n)\sin na.$$

These results are true when $a = \frac{1}{2}\pi$ if $n < 1$.

Deduce that

$$\int_0^\infty \cos(y^2)\,dy = \int_0^\infty \sin(y^2)\,dy = \left(\tfrac{1}{8}\pi\right)^{\frac{1}{2}}. \qquad \text{(Euler.)}$$

44. The contour C starts from a point R on the real axis, encircles the origin once counterclockwise and returns to R. By deforming the contour into two straight lines and a circle of radius δ (like the figure of § 29 with the large circle omitted), and making $\delta \to 0$, shew that, if $\xi > 0$, (where $\zeta = \xi + i\eta$), and $-\pi \leqslant \arg(-z) \leqslant \pi$ on C, then

$$\lim_{R\to\infty} \int_C (-z)^{\zeta-1} e^{-z}\,dz = -2i\sin(\pi\zeta)\,.\,\Gamma(\zeta).$$

(Hankel, *Math. Ann.* Bd. I.)

45. If $\Gamma(\zeta)$ be defined when $\xi \leqslant 0$ by means of the relation $\Gamma(\zeta+1) = \zeta\Gamma(\zeta)$, prove by integrating by parts that the equation of Example 44 is true for all values of ζ.

46. By taking a parabola, whose focus is at 0, as contour, shew that, if $a > 0$, then

$$\Gamma(\zeta) = \frac{2a^{\zeta} e^{a}}{\sin \pi\zeta} \int_0^{\infty} e^{-at^2} (1+t^2)^{\zeta - \frac{1}{2}} \cos\{2at + (2\zeta - 1) \operatorname{arc} \tan t\}\, dt$$

(Bourguet.)

47. Assuming Stirling's formula[10], namely that

$$\{\log \Gamma(z+1) - (z+\tfrac{1}{2}) \log z + z - \tfrac{1}{2} \log 2\pi\} \to 0$$

uniformly as $|z| \to \infty$, when $-\pi + \delta < \arg z < \pi - \delta$, and δ is any positive constant, shew that, if $-\frac{1}{2}\pi < \arg(-\zeta) < \frac{1}{2}\pi$, then

$$\frac{1}{2\pi i} \int_{-a-\infty i}^{-a+\infty i} \Gamma(-z)(-\zeta)^z\, dz = -\frac{1}{2\pi i} \int_{-a-\infty i}^{-a+\infty i} \frac{(-\zeta)^z}{\Gamma(z+1)} \frac{\pi\, dz}{\sin \pi z} = e^{\zeta},$$

where $a > 0$, and the path of integration is a straight line. (The expressions may be shewn to be the sum of the residues of the second integrand at its poles on the right of the path of integration.)

(Mellin, *Acta Soc. Fennicae*, vol. xx.)

48. Let C be a closed contour, and let

$$f(z) = \prod_{r=1}^{m} (z - a_r)^{n_r} \phi(z)$$

where the points a_r are inside C, the numbers n_r are integers (positive or negative), while $\phi(z)$ is analytic on and inside C and has no zeros on or inside C. Shew that, if $f'(z)$ be the derivate of $f(z)$, then

$$\frac{1}{2\pi i} \int_C \frac{f'(z)}{f(z)}\, dz = \sum_{r=1}^{m} n_r. \qquad \text{(Cauchy.)}$$

49. By taking the contour C of Example 48 to be a circle of radius R, and making $R \to \infty$, shew that a polynomial of degree n has n roots. (Cauchy.)

50. With the notation of Example 48, shew that, if $\psi(z)$ be analytic on and inside C, then

$$\frac{1}{2\pi i} \int_C \psi(z) \frac{f'(z)}{f(z)}\, dz = \sum_{r=1}^{m} n_r \psi(a_r). \qquad \text{(Cauchy.)}$$

[10] Stieltjes, *Liouville's Journal*, t. IV.

CHAPTER VII

EXPANSIONS IN SERIES

§ 34. Taylor's Theorem.—§ 35. Laurent's Theorem.

34. Taylor's Theorem. *Let $f(z)$ be a function of z which is analytic at all points inside a circle of radius r whose centre is the point whose complex coordinate is a. Let ζ be any point* inside *this circle.*

Then $f(\zeta)$ can be expanded into the convergent series:

$$f(\zeta)=f(a)+(\zeta-a)f'(a)+\frac{1}{2\,!}(\zeta-a)^2f''(a)+\ldots+\frac{1}{m\,!}(\zeta-a)^m f^{(m)}(a)+\ldots,$$

where $f^{(m)}(a)$ denotes $\dfrac{d^m f(a)}{da^m}$.

Let $|\zeta-a|=\theta r$, so that $0 \leqslant \theta < 1$.

Let C be the contour formed by the circle $|z-a|=\theta' r$, where $\theta<\theta'<1$; let $\theta/\theta'=\theta_1$, so that $|\zeta-a| \div |z-a|=\theta_1<1$; then by § 21,

$$f(\zeta)=\frac{1}{2\pi i}\int_C \frac{f(z)}{z-\zeta}\,dz$$

$$=\frac{1}{2\pi i}\int_C \frac{f(z)}{z-a}\left(1-\frac{\zeta-a}{z-a}\right)^{-1}dz$$

$$=\frac{1}{2\pi i}\int_C \frac{f(z)}{z-a}\left(1+\frac{\zeta-a}{z-a}+\frac{(\zeta-a)^2}{(z-a)^2}+\ldots+\frac{(\zeta-a)^n}{(z-a)^n}+\frac{(\zeta-a)^{n+1}}{(z-a)^n(z-\zeta)}\right)dz$$

$$=\frac{1}{2\pi i}\sum_{m=0}^{n}\int_C \frac{f(z)\,dz}{(z-a)^{m+1}}(\zeta-a)^m+\frac{1}{2\pi i}\int_C \frac{f(z)(\zeta-a)^{n+1}}{(z-a)^{n+1}(z-\zeta)}\,dz$$

$$=\sum_{m=0}^{n}\frac{a_m}{m\,!}(\zeta-a)^m+\frac{1}{2\pi i}\int_C \frac{f(z)(\zeta-a)^{n+1}}{(z-a)^{n+1}(z-\zeta)}\,dz,$$

where
$$a_m=\frac{m\,!}{2\pi i}\int_C \frac{f(z)\,dz}{(z-a)^{m+1}}$$

$$=f^{(m)}(a), \text{ by § 22.}$$

On C, $|f(z)|$ does not exceed some fixed[1] number K, since $f(z)$ is analytic and, *a fortiori*, continuous on C.

Therefore

$$\left|\frac{1}{2\pi i}\int_C \frac{f(z)(\zeta-a)^{n+1}}{(z-a)^{n+1}(z-\zeta)}dz\right| \leqslant \frac{1}{2\pi}\int_C \left|\frac{f(z)(\zeta-a)^{n+1}}{(z-a)^{n+1}(z-\zeta)}dz\right|$$

$$< \frac{1}{2\pi}\int_C \frac{K\theta_1^{n+1}}{(1-\theta_1)r}|dz|$$

$$< K\theta_1^{n+1}/(1-\theta_1),$$

since $$|z-\zeta| = |(z-a)-(\zeta-a)| \geqslant |z-a|-|\zeta-a|.$$

Now $\lim_{n\to\infty} K\theta_1^{n+1}/(1-\theta_1) = 0$, since $0 \leqslant \theta_1 < 1$; and therefore

$$\lim_{n\to\infty}\frac{1}{2\pi i}\int_C \frac{f(z)(\zeta-a)^{n+1}}{(z-a)^{n+1}(z-\zeta)}dz = 0;$$

consequently $f(\zeta) = \lim_{n\to\infty}\sum_{m=0}^{n}\frac{a_m}{m!}(\zeta-a)^m$; since this limit exists it follows that the series $\sum_{m=0}^{\infty}\frac{a_m}{m!}(\zeta-a)^m$ is convergent; and it has therefore been shewn that $f(\zeta)$ can be expanded into the convergent series:

$$f(\zeta) = a_0 + \frac{a_1}{1!}(\zeta-a) + \frac{a_2}{2!}(\zeta-a)^2 + \ldots + \frac{a_m}{m!}(\zeta-a)^m + \ldots$$

where $$a_m = f^{(m)}(a) = \frac{m!}{2\pi i}\int_C \frac{f(z)\,dz}{(z-a)^{m+1}}.$$

35. Laurent's Theorem. *Let $f(z)$ be a function of z which is analytic and one-valued at all points inside the region bounded by two oriented concentric circles (Γ, Γ'), centre a, radii r_1, r_1' where $r_1' < r_1$.*

Let ζ be any point inside Γ and outside Γ'; then $f(\zeta)$ can be expressed as the sum of two convergent series:

$$f(\zeta) = a_0 + a_1(\zeta-a) + a_2(\zeta-a)^2 + \ldots + a_m(\zeta-a)^m + \ldots$$
$$+ b_1(\zeta-a)^{-1} + b_2(\zeta-a)^{-2} + \ldots + b_m(\zeta-a)^{-m} + \ldots,$$

where $$a_m = \frac{1}{2\pi i}\int_C \frac{f(z)\,dz}{(z-a)^{m+1}},\quad b_m = \frac{1}{2\pi i}\int_{C'} (z-a)^{m-1}f(z)\,dz;$$

the circles C, C' are concentric with the circles Γ, Γ' and are of radii r, r' such that $r_1' < r' < |\zeta - a| < r < r_1$.

[1] See note 4 on p. 50.

Draw a diameter $ABCD$ of the circles C, C', not passing through ζ. Let C_1, C_1' be the semicircles on one side of this diameter, while C_2, C_2' are the semicircles on the other side.

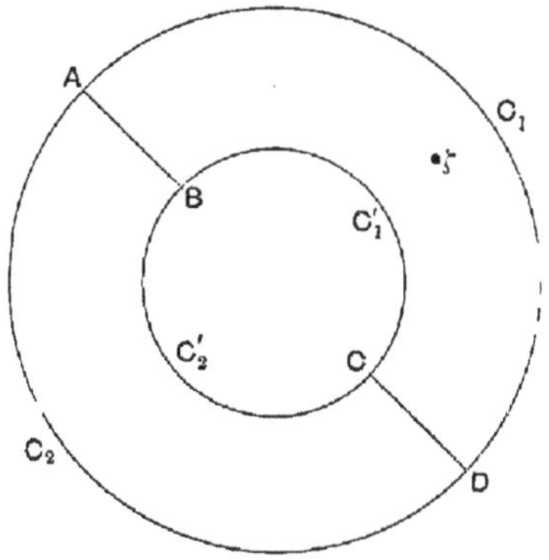

Then C_1, AB, C_1', CD can be oriented to form a contour Γ_1; and C_2, DC, C_2', BA can be oriented to form a contour Γ_2; it is easily seen that AB, BA have opposite orientations in the two contours, as do CD, DC; C_1, C_2 have the same orientation as C, C_1', C_2' have opposite orientations to C'; and $f(z)$ is analytic in the closed regions formed by Γ_1, Γ_2 and their interiors

Therefore $$\int_{\Gamma_1} \frac{f(z)}{z-\zeta}dz + \int_{\Gamma_2} \frac{f(z)}{z-\zeta}dz = \int_C \frac{f(z)}{z-\zeta}dz - \int_{C'} \frac{f(z)}{z-\zeta}dz;$$

the integrals along BA, AB cancel, and so do those along CD, DC.

But, by § 21, $$\int_{\Gamma_1} \frac{f(z)}{z-\zeta}dz + \int_{\Gamma_2} \frac{f(z)}{z-\zeta}dz = 2\pi i f(\zeta);$$

for ζ is inside one of the contours Γ_1, Γ_2 and outside the other.

Therefore $$f(\zeta) = \frac{1}{2\pi i}\int_C \frac{f(z)}{z-\zeta}dz - \frac{1}{2\pi i}\int_{C'} \frac{f(z)}{z-\zeta}dz.$$

But $$\frac{1}{2\pi i}\int_C \frac{f(z)}{z-\zeta}dz = \frac{1}{2\pi i}\int_C f(z) \sum_{m=0}^{n} \frac{(\zeta-a)^m}{(z-a)^{m+1}}dz$$
$$+ \frac{1}{2\pi i}\int_C f(z) \frac{(\zeta-a)^{n+1}}{(z-a)^{n+1}(z-\zeta)}dz$$
$$= \sum_{m=0}^{n} a_m(\zeta-a)^m + \frac{1}{2\pi i}\int_C \frac{f(z)(\zeta-a)^{n+1}}{(z-a)^{n+1}(z-\zeta)}dz.$$

By the arguments of § 34, it may be shewn that the last integral tends to zero as $n \to \infty$; so that $\frac{1}{2\pi i}\int_C \frac{f(z)}{z-\zeta}\,dz$ can be expanded into the convergent series $\sum_{m=0}^{\infty} a_m(\zeta-a)^m$.

In like manner

$$\begin{aligned} -\frac{1}{2\pi i}\int_{C'} \frac{f(z)}{z-\zeta}\,dz &= \frac{1}{2\pi i}\int_{C'} \frac{f(z)}{\zeta-a}\left(1-\frac{z-a}{\zeta-a}\right)^{-1} dz \\ &= \frac{1}{2\pi i}\int_{C'} f(z) \sum_{m=1}^{n} \frac{(z-a)^{m-1}}{(\zeta-a)^m}\,dz \\ &\quad + \frac{1}{2\pi i}\int_{C'} \frac{f(z)(z-a)^n}{(\zeta-a)^n(\zeta-z)}\,dz \\ &= \sum_{m=1}^{n} b_m(z-a)^{-m} + \frac{1}{2\pi i}\int_{C'} \frac{f(z)(z-a)^n}{(\zeta-a)^n(\zeta-z)}\,dz. \end{aligned}$$

Since, on C', $\left|\frac{z-a}{\zeta-a}\right| < 1$, the arguments of § 34 can be applied to shew that the last integral tends to zero as $n \to \infty$; so that

$$-\frac{1}{2\pi i}\int_{C'} \frac{f(z)}{z-\zeta}\,dz$$

can be expanded into the convergent series $\sum_{m=1}^{\infty} b_m(z-a)^{-m}$; that is to say

$$f(\zeta) = \sum_{m=0}^{\infty} a_m(\zeta-a)^m + \sum_{m=1}^{\infty} b_m(\zeta-a)^{-m},$$

each of the series being convergent.

CHAPTER VIII

HISTORICAL SUMMARY

§ 36. Definitions of analytic functions.—§ 37. Proofs of Cauchy's theorem.

36. The earliest suggestion of the theorem to which Cauchy's name has been given is contained in a letter[1] from Gauss to Bessel dated Dec. 18, 1811; in this letter Gauss points out that the value of $\int x^{-1}\,dx$ taken along a complex path depends on the path of integration. The earliest investigation of Cauchy on the subject is contained in a memoir[2] dated 1814, and a formal proof of the complete theorem is given in a memoir[3] published in 1825.

The proof contained in this memoir consists in proving that the variation of $\int_{(AB)} f(z)\,dz$, when the path of integration undergoes a small variation (the end-points remaining fixed), is zero, provided that $f(z)$ has a unique continuous differential coefficient at all points on the path AB.

The following is a summary of the various assumptions on which proofs of Cauchy's theorem have been based:

(i) The hypothesis of Goursat[4]: $f'(z)$ exists at all points within and on C.

(ii) The hypothesis of Cauchy[5]: $f'(z)$ exists and is *continuous*.

[1] *Briefwechsel zwischen Gauss und Bessel* (1880), pp. 156–157.

[2] *Oeuvres complètes*, sér. I, t. 1, p. 402 *et seq.*

[3] *Mémoire sur les intégrales définies prises entre des limites imaginaires.* References to Cauchy's subsequent researches are given by Lindelöf, *Calcul de Résidus.*

[4] *Transactions of the American Mathematical Society*, vol. I (1900), pp. 14–16.

[5] See the memoir cited above.

(iii) An hypothesis equivalent to the last is: $f(z)$ is uniformly differentiable; i.e., when ϵ is taken arbitrarily, then a positive δ, *independent* of z can be found such that whenever z and z' are on or inside C and $|z'-z| \leqslant \delta$, then

$$|f(z')-f(z)-(z'-z)f'(z)| \leqslant \epsilon|z'-z|.$$

In the language of Chapter II, this inequality enables us to take squares whose sides are not greater than $\delta/\sqrt{2}$ as 'suitable regions.'

(iv) The hypothesis of Riemann[6]: $f(z)=P+iQ$ where P, Q are real and have continuous derivates with respect to x and y such that

$$\frac{\partial P}{\partial x}=\frac{\partial Q}{\partial y}, \quad \frac{\partial Q}{\partial x}=-\frac{\partial P}{\partial y}.$$

These hypotheses are effectively equivalent, but, of course, (i) is the most natural starting-point of a development of the theory of functions on the lines laid down by Cauchy. It is easy to prove the equivalence[7] of (ii), (iii) and (iv), but attempts at deducing any one of these three from (i), except by means of Cauchy's theorem and the results of §§ 21–22, have not been successful; however, it is easy to deduce from § 22, by using the expression for $f'(z)$ as a contour integral, that, if (i) is assumed, then (ii) is true in the interior of C.

The definition of Weierstrass is that an analytic function $f(z)$ is such that it can be expanded into a Taylor's series in powers of $z-a$ where a is a point inside C. This hypothesis is simple and fundamental in the Weierstrassian theory of functions, in which Cauchy's theorem appears merely incidentally.

37. A proof of Cauchy's theorem, based on hypothesis (i), requires Goursat's lemma (which is a special case of the Heine-Borel theorem) or its equivalent; the apparent exception, a proof due to Moore[8], employs, in the course of the proof, arguments similar to those by which Goursat's lemma is proved.

The hypotheses (ii) and (iii) are such as to make it easy to divide C and its interior into suitable regions.

The various methods of proof of the theorem are the following:

(i) Goursat's proof, first published in 1884 [this, in its earliest form[9], employs hypothesis (iii)], is essentially that given in this work.

[6] *Oeuvres mathématiques* (1898), Dissertation inaugurale (1851).

[7] The equivalence of (ii) and (iii) has been proved in § 20.

[8] *Transactions of the American Mathematical Society*, vol. I (1900), pp. 499–506.

[9] *Acta Mathematica*, vol. IV, pp. 197–200; see also his *Cours d'Analyse*, t. II.

(ii) Cauchy's proof has already been described.

(iii) Riemann's proof[10] consists in transforming

$$\int(Pdx - Qdy) + i\int(Qdx + Pdy)$$

into a double integral, by using Stokes' theorem.

(iv) Moore's proof consists in assuming that the integral taken round the sides of a square is *not* zero, but has modulus η_0; the square is divided into four equal squares, and the modulus of the integral along at least one of these must be $\geqslant \frac{1}{4}\eta_0$; the process of subdividing squares is continued, giving rise to at least one limiting point ζ inside every square S_ν of a sequence such that the modulus of the integral along S_ν is not less than $\eta_0/4^\nu$. Assuming that $f(z)$ has a derivate at ζ it is proved that it is possible to find ν_0 such that, when $\nu > \nu_0$, the modulus of the integral along S_ν is less than $\eta_0/4^\nu$. This is the contradiction needed to complete the proof of the theorem. The deduction of the theorem for a closed contour, not a square, may then be obtained by the methods given above in § 17.

Finally, it should be mentioned that, although the use of Cauchy's theorem may afford the simplest method of evaluating a definite integral, the result can always be obtained by other methods; thus a *direct* use of Cauchy's theorem can always be avoided, if desired, by transforming the contour integral into a double integral as in Riemann's proof. Further, Cauchy's theorem cannot be employed to evaluate all definite integrals; thus $\int_0^\infty e^{-x^2}\,dx$ has not been evaluated except by other methods.

[10] See the dissertation cited above.

CPSIA information can be obtained
at www.ICGtesting.com
Printed in the USA
BVHW072244040720
582971BV00003B/261

9 781375 422697